D1243712

AN ANNOTATED TIMELINE OF OPERATIONS RESEARCH

An Informal History

Recent titles in the
INTERNATIONAL SERIES IN
OPERATIONS RESEARCH & MANAGEMENT SCIENCE
Frederick S. Hillier, Series Editor, *Stanford University*

Ramík, J. & Vlach, M. / *GENERALIZED CONCAVITY IN FUZZY OPTIMIZATION AND DECISION ANALYSIS*
Song, J. & Yao, D. / *SUPPLY CHAIN STRUCTURES: Coordination, Information and Optimization*
Kozan, E. & Ohuchi, A. / *OPERATIONS RESEARCH/MANAGEMENT SCIENCE AT WORK*
Bouyssou et al. / *AIDING DECISIONS WITH MULTIPLE CRITERIA: Essays in Honor of Bernard Roy*
Cox, Louis Anthony, Jr. / *RISK ANALYSIS: Foundations, Models and Methods*
Dror, M., L'Ecuyer, P. & Szidarovszky, F. /*MODELING UNCERTAINTY: An Examination of Stochastic Theory, Methods, and Applications*
Dokuchaev, N. / *DYNAMIC PORTFOLIO STRATEGIES: Quantitative Methods and Empirical Rules for Incomplete Information*
Sarker, R., Mohammadian, M. & Yao, X. / *EVOLUTIONARY OPTIMIZATION*
Demeulemeester, R. & Herroelen, W. / *PROJECT SCHEDULING: A Research Handbook*
Gazis, D.C. / *TRAFFIC THEORY*
Zhu, J. / *QUANTITATIVE MODELS FOR PERFORMANCE EVALUATION AND BENCHMARKING*
Ehrgott, M. & Gandibleux, X. / *MULTIPLE CRITERIA OPTIMIZATION: State of the Art Annotated Bibliographical Surveys*
Bienstock, D. / *Potential Function Methods for Approx. Solving Linear Programming Problems*
Matsatsinis, N.F. & Siskos, Y. / *INTELLIGENT SUPPORT SYSTEMS FOR MARKETING DECISIONS*
Alpern, S. & Gal, S. / *THE THEORY OF SEARCH GAMES AND RENDEZVOUS*
Hall, R.W. / *HANDBOOK OF TRANSPORTATION SCIENCE – 2nd Ed.*
Glover, F. & Kochenberger, G.A. / *HANDBOOK OF METAHEURISTICS*
Graves, S.B. & Ringuest, J.L. / *MODELS AND METHODS FOR PROJECT SELECTION: Concepts from Management Science, Finance and Information Technology*
Hassin, R. & Haviv, M. / *TO QUEUE OR NOT TO QUEUE: Equilibrium Behavior in Queueing Systems*
Gershwin, S.B. et al. / *ANALYSIS & MODELING OF MANUFACTURING SYSTEMS*
Maros, I. / *COMPUTATIONAL TECHNIQUES OF THE SIMPLEX METHOD*
Harrison, T., Lee, H. & Neale, J. / *THE PRACTICE OF SUPPLY CHAIN MANAGEMENT: Where Theory And Application Converge*
Shanthikumar, J.G., Yao, D. & Zijm, W.H. / *STOCHASTIC MODELING AND OPTIMIZATION OF MANUFACTURING SYSTEMS AND SUPPLY CHAINS*
Nabrzyski, J., Schopf, J.M. & Węglarz, J. / *GRID RESOURCE MANAGEMENT: State of the Art and Future Trends*
Thissen, W.A.H. & Herder, P.M. / *CRITICAL INFRASTRUCTURES: State of the Art in Research and Application*
Carlsson, C., Fedrizzi, M. & Fuller, R. / *FUZZY LOGIC IN MANAGEMENT*
Soyer, R., Mazzuchi, T.A. & Singpurwalla, N.D. / *MATHEMATICAL RELIABILITY: An Expository Perspective*
Talluri, K. & van Ryzin, G. / *THE THEORY AND PRACTICE OF REVENUE MANAGEMENT*
Kavadias, S. & Loch, C.H. / *PROJECT SELECTION UNDER UNCERTAINTY: Dynamically Allocating Resources to Maximize Value*
Sainfort, F., Brandeau, M.L. & Pierskalla, W.P. / *HANDBOOK OF OPERATIONS RESEARCH AND HEALTH CARE: Methods and Applications*
Cooper, W.W., Seiford, L.M. & Zhu, J. / *HANDBOOK OF DATA ENVELOPMENT ANALYSIS: Models and Methods*
Sherbrooke, C.C. / *OPTIMAL INVENTORY MODELING OF SYSTEMS: Multi-Echelon Techniques – 2nd Ed.*
Chu, S.-C., Leung, L.C., Hui, Y.V. & Cheung, W. / *4th PARTY CYBER LOGISTICS FOR AIR CARGO*
Simchi-Levi, Wu, Shen / *HANDBOOK OF QUANTITATIVE SUPPLY CHAIN ANALYSIS: Modeling in the E-Business Era*

***** A list of the early publications in the series is at the end of the book *****

An Annotated Timeline of Operations Research: An Informal History

Saul I. Gass

Arjang A. Assad

Robert H. Smith School of Business
University of Maryland, College Park

Saul I. Gass
University of Maryland
USA

Arjang A. Assad
University of Maryland
USA

Library of Congress Cataloging-in-Publication Data

Gass, Saul I.
An annotated timeline of operations research : an informal history / Saul I. Gass, Arjang A. Assad.
p. cm. – (International series in operations research & management science; 75)
Includes bibliographical references and index.
ISBN 1-4020-8112-X --- ISBN 1-4020-8116-2 (pbk.)
1. Operations research--History. I. Assad, A. (Arjang). II. Title. II. Series.

T57.6.G37 2004
658.4'034'09—dc22

2004054946

Printed in the United States of America.

9 8 7 6 5 4 3 2 1 SPIN 11041627

springeronline.com

To Arianna,

who brings joy to all,
especially to her Granddad.

To my mother, Derakhshandeh,

the source of my informal history –
for her courage and patience.

Contents[1]

[1]The items in the *Annotated Timeline* have been divided into eight time-sequenced parts. Parts 1 and 2 (from 1564 to 1935) present the precursor scientific and related contributions that have influenced the subsequent development of operations research (OR). Parts 3 to 8 (from 1936 to 2004) describe the beginnings of OR and its evolution into a new science. They are so divided mainly for presentation purposes.

Preface

"What's past is prologue."

The Tempest, William Shakespeare, Act II, Scene I

Dictionary definitions of a scientific field are usually clear, concise and succinct. Physics: "The science of matter and energy and of interactions between the two;" Economics: "The science that deals with the production, distribution, and consumption of commodities;" Operations Research (OR): "Mathematical or scientific analysis of the systematic efficiency and performance of manpower, machinery, equipment, and policies used in a governmental, military or commercial operation." OR is not a natural science. OR is not a social science. As implied by its dictionary definition, OR's distinguishing characteristic is that OR applies its scientific and technological base to resolving problems in which the human element is an active participant. As such, OR is the science of decision making, the science of choice.

What were the beginnings of OR? Decision making started with Adam and Eve. There are apocryphal legends that claim OR stems from biblical times – how Joseph aided Pharaoh and the Egyptians to live through seven fat years followed by seven lean years by the application of "lean-year" programming. The Roman poet Virgil recounts in the Aeneid the tale of Dido, the Queen of Carthage, who determined the maximum amount of land that "could be encircled by a bull's hide." The mathematicians of the seventeenth and eighteenth centuries developed the powerful methods of the calculus and calculus of variations and applied them to a wide range of mathematical and physical optimization problems. In the same historical period, the basic laws of probability emerged in mathematical form for the first time and provided a basis for making decisions under uncertainty.

But what events have combined to form OR, the science that aids in the resolution of human decision-making problems? As with any scientific field, OR has its own "pre-history," comprised of a collection of events, people, ideas, and methods that contributed to the study of decision-making even before the official birth of OR. Accordingly, the entries in *An Annotated Timeline of Operations Research* try to capture some of the key events of this pre-history.

Many of the early operations researchers were trained as mathematicians, statisticians and physicists; some came from quite unrelated fields such as chemistry, law, history, and psychology. The early successes of embryonic OR prior to and during World War II illustrate the essential feature that helped to establish OR: bright, well-trained, curious, motivated people, assigned to unfamiliar and difficult problem settings, most often produce improved solutions. A corollary is that a new look, a new analysis, using methods foreign to the original problem environment can often lead to new insights and new solutions. We were fortunate to have leaders who recognized this fact; scientists such as Patrick M. S. Blackett and Philip M. Morse and their military coworkers. They were not afraid to challenge the well-intentioned in-place bureaucracy in their search to improve both old and new military operations. The urgency of World War II allowed this novel approach to prove itself. And, the foresight of these early researchers led to the successful transfer of OR to post-war commerce and industry. Today, those who practice or do research in OR can enter the field through various educational and career paths, although the mathematical language of OR favors disciplines that provide training in the use of mathematics.

Blackett and Morse brought the scientific method to the study of operational problems in a manner much different from the earlier scientific management studies of Fredrick Taylor and Frank and Lillian Gilbreth. The latter worked a problem over – collected and analyzed related data, trying new approaches such as a shovel design (Taylor) or a new sequence for laying bricks (F. Gilbreth), and, in general, were able to lower costs and achieve work efficiencies. From today's perspective, what was missing from their work was (1) the OR emphasis on developing theories about the process under study, that is modeling, with the model(s) being the scientist's experimental laboratory where alternative solutions are evaluated against single or multiple measures of effectiveness, combined with (2) the OR philosophy of trying to take an holistic view of the system under study. It must be noted, however, that much of early OR did not fit this pattern: actual radar field experiments were conducted by the British for locating aircraft, as well as real-time deployment of fighter aircraft experiments; new aircraft bombing patterns were tried and measured against German targets; new settings for submarine depth charges were proven in the field. But such ideas were based on analyses of past data and evaluated by studies of the "experimental" field trials. Some models and modeling did come into play: submarine-convoy tactics were gamed on a table, and new bombing strategies evolved from statistical models of past bomb dispersion patterns.

The history of OR during World War II has been told in a number of papers and books (many cited in the *Annotated Timeline*). What has not been told in any depth is how OR moved from the classified confines of its military origins into being a new science. That story remains to be told. It is hidden in the many citations of the *Annotated Timeline*. It is clear, however, that the initial impetus was due to a few of the civilian and military OR veterans of World War II who believed that OR had value beyond the military. Post World War II we find: OR being enriched by new disciples from the academic and business communities; OR being broadened by new mathematical, statistical, and econometric ideas, as well as being influenced by other fields of human and industrial activities; OR techniques developed and extended by researchers and research centers; OR made doable and increasingly powerful through the advent of the digital computer; OR being formalized and modified by new academic programs; OR going world-wide by the formation of country-based and international professional organizations; OR being supported by research journals established by both professional organizations and scientific publishers; and OR being sustained by a world-wide community of concerned practitioners and academics who volunteer to serve professional organizations, work in editorial capacities for journals, and organize meetings that help to announce new technical advances and applications.

Although our *Annotated Timeline* starts in 1564, the scope of what is today's OR is encompassed by a very short time period – just over three score years measured from 1936. In charting the timeline of OR, beginning with World War II, we are fortunate in having access to a full and detailed trail of books, journal articles, conference proceedings, and OR people and others whose mathematical, statistical, econometric, and computational discoveries have formed OR. The *Annotated Timeline* basically stops in the 1990s, although there are a few items cited in that time period and beyond. We felt too close to recent events to evaluate their historical importance. Future developments will help us decide what should be included in succeeding editions of the *Annotated Timeline*.

We believe that the *Annotated Timeline* recounts how the methodology of OR developed in some detail. In contrast, the *Annotated Timeline* gives only partial coverage to the practical side of OR. We felt, from an editorial point-of-view, that it would be counterproductive to note a succession of applications. Further, the telling would be incomplete: unlike academics, practitioners tend not to publish accounts of their activities and are often constrained from publishing for proprietary reasons. Thus, the *Annotated Timeline* gives the reader a restricted view of the practice of OR. To counter this, we suggest that the interested reader review the past volumes of the journal *INTERFACES*, especially the issues that contain the well-written OR practice papers that describe the work of the Edelman Prize competition finalists. Collectively, they tell an amazing story: how the wide-ranging practical application of OR has furthered the advancement of commerce, industry, government, and the military, as no other science has done in the past. (Further sources of applications are *Excellence in Management Science Practice: A Readings Book*, A. A. Assad, E. A. Wasil, G. L. Lilien, Prentice-Hall, Englewood Cliffs, 1992, and *Encyclopedia of Operations Research and Management Science*, 2nd edition, S. I. Gass, C. M. Harris, Kluwer Academic Publishers, Boston, 2001.)

In selecting and developing a timeline entry, we had several criteria in mind: we wanted it to be historically correct, offer the reader a concise explanation of the event under discussion, and to be a source document in the sense that references for an item would enable the reader to obtain more relevant information, especially if these references contained significant historical information. Not all items can be neatly pegged to a single date, and the exact beginnings of some ideas or techniques are unclear. We most often cite the year in which related material was first published. In some instances, however, we used an earlier year if we had confirming information. For many entries, we had to face the conflicting requirements imposed between the timeline and narrative formats. A timeline disperses related events along the chronological line by specific dates, while annotations tend to cluster a succession of related events into the same entry. We generally used the earliest date to place the item on the timeline, and discuss subsequent developments in the annotation for that entry. Some items, however, evolved over time and required multiple entries. We have tried to be as complete and correct as possible with respect to originators and authorship. We also cite a number of books and papers, all of which have influenced the development of OR and helped to educate the first generations of OR academics and practitioners.

No timeline constrained to a reasonable length can claim to be complete. Even the totality of entries in this *Annotated Timeline* does not provide a panoramic view of the field. Entries were selected for their historical import, with the choices clearly biased towards pioneering works or landmark developments. Sometimes, an entry was included as it related a conceptual or mathematical advance or told an interesting historical tale.

OR is a rich field that draws upon several different disciplines and methodologies. This makes the creation of a timeline more challenging. How does one negotiate the boundaries between OR, economics, industrial engineering, applied mathematics, statistics, or computer science, not to mention such functional areas as operations management or marketing? While we had to make pragmatic choices, one entry at a time, we were conscious that our choices reflect our answer to the basic question of "What is OR?" We recognize that the answer to this question and the drawing of the boundaries of OR varies depending on the background and interests of the respondent.

We wish to thank the many people who were kind enough to suggest items, offer corrections, and were supportive of our work. We made many inquiries of friends and associates around the world. All have been exceptionally responsive to our request for information, many times without knowing why we asked such questions as "What is the first name of so-and-so?" and "When did this or that begin?" Any errors and omissions are, of course, our responsibility. We trust the reader will bring the details of any omission to our attention. We look forward to including such additional timeline entries – those that we missed and those yet to be born – in future editions of the *Annotated Timeline*. In anticipation, we await, with thanks, comments and suggestions from the reader.

We are especially appreciative of Kluwer Academic Publisher's administrative and production staffs for their truly professional approach to the development and production of the *Annotated Timeline*. In particular, we wish to acknowledge the support and cooperation of editor Gary Folven, production editor Katie Costello, and series editor Fred Hillier.

Saul I. Gass

Arjang A. Assad

We wish to acknowledge and thank the many individuals who sent us pictures and gave us permission to use them; they are too many to list. We also wish to thank the following organizations: The Nobel Foundation, Institute of Electrical and Electronics Engineers, American Economic Review, The Econometric Society, American Statistical Association, The Library of Congress, The RAND Corporation, Harvard University Photo Services, The W. Edwards Deming Institute, MIT Press.

A note on how books and papers are cited: (1) Books called out explicitly as timeline items are given by year of publication, title (bold type) in italics, author(s), publisher, city; (2) Books as references for a timeline item are given by title in italics, author(s), publisher, city, year of publication; (3) Papers called out explicitly as timeline items are given by year of publication, title (bold type) in quotes, author(s), journal in italics, volume number, issue number, page numbers; (4) Papers as references for a timeline item are given by title in quotes, author(s), journal in italics, volume number, issue number, year of publication, page numbers [for (3) and (4), if there is only one number after the journal name and before the year, it is the volume number].

1

Operations research precursors from 1564 to 1873

1564 ***Liber de Ludo Aleae (The Book on Games of Chance)***, Girolamo Cardano, pp. 181–243 in *Cardano: The Gambling Scholar,* Oystein Ore, Dover Publications, New York, 1965

Girolamo Cardano, Milanese physician, mathematician and gambler, is often cited as the first mathematician to study gambling. His book, *Liber de Ludo Aleae* (The Book on Games of Chance), is devoted to the practical and theoretical aspects of gambling. Cardano computes chance as the ratio between the number of favorable outcomes to the total number of outcomes, assuming outcomes are equally likely. The *Book* remained unpublished until 1663 by which time his results were superseded by the work of Blaise Pascal and Pierre de Fermat in 1654. Franklin (2001) traces the history of rational methods for dealing with risk to classical and medieval ages. [*A History of Probability and Statistics and Their Applications Before 1750*, A. Hald, John Wiley & Sons, New York, 1990; *The Science of Conjecture: Evidence and Probability before Pascal*, J. Franklin, The John Hopkins University Press, Baltimore, 2001]

Charter member of gambler's anonymous:

Cardano wrote in his autobiography that he had "an immoderate devotion to table games and dice. During many years – for more than forty years at the chess boards and twenty-five years of gambling – I have played not off and on but, as I am ashamed to say, every day." (Hald, 1990)

Girolamo Cardano
1501–1576

1654 Expected value

The French mathematician, Blaise Pascal, described how to compute the expected value of a gamble. In his letter of July 29, 1654 to Pierre de Fermat, Pascal used the key idea of equating the value of the game to its mathematical expectation computed as the probability of a win multiplied by the gain of the gamble. Jakob Bernoulli I called this "the fundamental principle of the whole art" in his *Ars Conjectandi* (1713). [*Mathematics: Queen and Servant of Science*, E. T. Bell, McGraw-Hill, New York, 1951; *A History of Probability and Statistics and Their Applications Before 1950*, A. Hald, John Wiley & Sons, New York, 1990]

Pascal's wager:

Pascal used his concept of mathematical expectation to resolve what is known as "Pascal's wager": Since eternal happiness is infinite, and even if the probability of winning eternal happiness by leading a religious life is very small, the expectation is infinite and, thus, it would pay to lead a "godly, righteous, and sober life." Pascal took his own advice.

Blaise Pascal
1623–1662

1654 The division of stakes: The problem of points

Two players, A and B, agree to play a series of fair games until one of them has won a specified number g of games. If the play is stopped prematurely when A has won r games and B has won s games (with r and s both smaller than g), how should the stakes be divided between A and B? This division problem (or the problem of points) was discussed and analyzed by various individuals since 1400. Girolamo Cardano gave one of the first correct partial solutions of this problem in 1539. The full solution, which laid the foundation of probability theory, was stated in the famous correspondence between Blaise Pascal and Pierre de Fermat in 1654. Pascal used recursion relations to solve the problem of points while Fermat enumerated the various possible combinations. Pascal also communicated a version of the gambler's ruin problem to Fermat, where the players had unequal probabilities of winning each game. [*A History of Probability and Statistics and Their Applications Before 1750*, A. Hald, John Wiley & Sons, New York, 1990; *The Science of Conjecture: Evidence and Probability Before Pascal*, J. Franklin, The John Hopkins University Press, Baltimore, 2001]

1657 *De Ratiociniis in Ludo Aleae* (*On Reckoning at Games of Chance*), Christiaan Huygens

Although more known for his work in constructing telescopes and inventing the pendulum clock, the Dutch scientist Christiaan Huygens wrote what is considered to be the first modern book on probability theory. It is noted for containing the formal definition of expectation and Huygens' recursive method for solving probability problems. Starting from an axiom on the fair value of a game, which Huygens called *expectatio*, the treatise states three theorems on expectations. Huygens uses these to solve several problems related to games of chance, some of which duplicate Pascal's work. Huygens had heard of Pascal's results but had not had the opportunity to meet him or examine his proofs. He therefore provided his own solutions and proofs. Later, Jakob Bernoulli I devoted the first part of his book *Ars Conjectandi* to an annotated version of Huygens' treatise. ["Huygens, Christiaan," H. Freudenthal, pp. 693–694 in *Encyclopedia of Statistical Sciences*, Vol. 6, S. Kotz, N. L. Johnson, editors, John Wiley & Sons, New York, 1985]

A best seller:

The Latin version of Huygen's book, published in September 1657, remained influential and was widely used for 50 years.

Christiaan Huygens
1629–1695

1662 Empirical probabilities for vital statistics

John Graunt, a tradesman from London, was the first English vital statistician. He used the data from bills of mortality to calculate empirical probabilities for such events as plague deaths, and rates of mortality from different diseases. In England, Bills of Mortality were printed in 1532 to record plague deaths, and weekly bills of christenings and burials started to appear in 1592. Graunt's book, *Natural and Political Observations on the Bills of Mortality*, appeared in 1662 and contained the first systematic attempt to extract reliable probabilities from bills of mortality. For instance, Graunt found that of 100 people born, 36 die before reaching the age of six, while seven survive to age 70. Graunt's calculations produced the first set of crude life tables. Graunt's book and the work of Edmund Halley on life tables (1693) mark the beginnings of actuarial science. De Moivre continued the analysis of annuities in his book *Annuities upon Lives* (1725). [*Games, Gods, and Gambling: A History of Probability and Statistical Ideas*, F. N. David, C. Griffin, London, 1962 (Dover reprint 1998); *Statisticians of the Centuries*, G. C. Heyde, E. Seneta, editors, Springer-Verlag, New York, 2001]

1665 Sir Isaac Newton

As with most scientific fields, OR has been influenced by the work of Sir Isaac Newton. In particular, two of Newton's fundamental mathematical discoveries stand out: finding roots of an equation and first order conditions for extrema. For equations, Newton developed an algorithm for finding an approximate solution (root) to the general equation $f(x) = 0$ by iterating the formula $x_{k+1} = x_k - f(x_k)/f'(x_k)$. Newton's Method can be used for finding the roots of a function of several variables, as well as the minimum of such functions. It has been adapted to solve nonlinear constrained optimization problems, with additional application to interior point methods for solving linear-programming problems. For a real-valued function $f(x)$, Newton gave $f'(x) = 0$ as the necessary condition for an extremum (maximum or minimum) of $f(x)$. About 35 years earlier, Fermat had implicitly made use of this condition when he solved for an extremum of $f(x)$ by setting $f(x)$ equal to $f(x+e)$ for a perturbation term e. Fermat, however, did not consider the notion of taking limits and the derivative was unknown to him. ["Fermat's methods of maxima and minima and of tangents: A reconstruction," P. Strømholm, *Archives for the History of Exact Sciences*, 5, 1968, 47–69; *The Mathematical Papers of Isaac Newton*, Vol. 3, D. T. Whiteside, editor, Cambridge University Press, Cambridge, 1970, 117–121; *The Historical Development of the Calculus*, C. H. Edwards, Jr., Springer-Verlag, New York, 1979; *Introduction to Numerical Analysis*, J. Stoer, R. Bulirsch, Springer-Verlag, New York, 1980; *Linear and Nonlinear Programming*, 2nd edition, D. G. Luenberger, Addison-Wesley, Reading, 1984; *Primal–Dual Interior-Point Methods*, S. J. Wright, SIAM, Philadelphia, 1997]

Go with the flow:

In his mathematical masterpiece on the calculus, *De Methodis Serierium et Fluxionum* (The Methods of Series and Fluxions), Newton stated: "When a quantity is greatest or least, at that moment its flow neither increases nor decreases: for if it increases, that proves that it was less and will at once be greater than it now is, and conversely so if it decreases. Therefore seek its fluxion... and set it equal to zero."

Isaac Newton
1642–1727

1713 The weak law of large numbers

In his book, *Ars Conjectandi*, Jakob Bernoulli I proved what is now known as Bernoulli's weak law of large numbers. He showed how to measure the closeness, in terms of a probability statement, between the mean of a random sample and the true unknown mean of the population as the sample size increases. Bernoulli was not just satisfied with the general result; he wanted to find the sample size that would achieve a desired closeness. As an illustrative example, Bernouilli could guarantee that with a probability of over $1000/1001$, a sample size of $N = 25,500$ would produce an observed relative frequency that fell within 1/50 of the true proportion of 30/50. [*The History of Statistics*, S. M. Stigler, Harvard University Press, Cambridge, 1986]

1713 St. Petersburg Paradox

In 1713, Nicolaus Bernoulli II posed five problems in probability to the French Mathematician Pierre Rémond de Montmort of which one was the following: "Peter tosses a coin and continues to do so until it should land 'heads' when it comes to the ground. He agrees to give Paul one ducat if he gets 'heads' on the very first throw, two ducats if he gets it on the second, four if on the third, eight if on the fourth, and so on, so that with each additional throw the number of ducats he must pay is doubled. Suppose we seek to determine the value of Paul's expectation." It is easy to show that the expectation is infinite; if that is the case, Paul should be willing to pay a reasonable amount to play the game. The question is "How much?" In answering this question twenty-five years later, Daniel Bernoulli, a cousin of Nicolaus, was the first to resolve such problems using the concept of (monetary) expected utility. The answer, according to Daniel Bernoulli is about 13 ducats. ["Specimen theoriae novae de mensura sortis," D. Bernoulli, *Commentarii Academiae Scientiarum Imperialis Petropolitanae*, Tomus V (Papers of the Imperial Academy of Sciences in Petersburg, Volume V), 1738, 175–192, English translation by L. Sommer, "Exposition of a new theory on the measurement of risk," D. Bernoulli, *Econometrica*, 22, 1954, 23–36; *Utility Theory: A Book of Readings*, A. N. Page, editor, John Wiley & Sons, New York, 1968; "The Saint Petersburg Paradox 1713–1937," G. Jorland, pp. 157–190 in *The Probabilistic Revolution, Vol. 1: Ideas in History*, L. Krüger, L. J. Daston, M. Heidelberger, editors, MIT Press, Cambridge, Mass., 1987; "The St. Petersburg Paradox" G. Shafer, pp. 865–870 in *Encyclopedia of Statistical Sciences*, Vol. 8, S. Kotz, N. L. Johnson, editors, John Wiley & Sons, New York, 1988]

Why a ducat?:

It is called the St. Petersburg Paradox as Daniel Bernoulli spent eight years in St. Petersburg and published an account in the *Proceedings of the St. Petersburg Academy of Science* (1738). In arriving at his answer of 13 ducats, Bernoulli assumed that the monetary gain after 24 successive wins, 166,777,216 ducats, represented the value he was willing to live with no matter how many heads came up in succession.

Daniel Bernoulli
1700–1782

1713 The earliest minimax solution to a game

James Waldegrave, Baron Waldegrave of Chewton, England, proposed a solution to the two-person version of the card game *Her* discussed by Nicolaus Bernoulli II and Pierre Rémond de Montmort in their correspondence. Waldegrave considered the problem of choosing a strategy that maximizes a player's probability of winning, no matter what strategy was used by the opponent. His result yielded what is now termed a minimax solution, a notion that forms the core of modern game theory. Waldegrave did not generalize the notion to other games; his minimax solution remained largely unnoticed. It was rediscovered by the statistician Ronald A. Fisher. [*A History of Probability and Statistics and Their Applications Before 1750*, A. Hald, John Wiley & Sons, New York, 1990; "The early history of the theory of strategic games from Waldegrave to Borel," R. W. Dimand, M. A. Dimand in *Toward a History of Game Theory*, E. R. Weintraub, editor, Duke University Press, Durham, 1992]

> *The game of Her:*
> Two players, A and B, draw cards in succession from a pack of 52 cards with cards numbered from 1 to 13 in four suits. A can compel B to exchange cards unless B has a 13. If B is not content with B's original card, or with the card held after the exchange with A, B can draw randomly from the remaining 50 cards, but if this card is a 13, B is not allowed to change cards. A and B then compare cards and the player with the higher card wins. B wins if the cards have equal value.

1715 Taylor series

Early in the eighteenth century, mathematicians realized that the expansions of various elementary transcendental functions were special cases of the general series now known as Taylor series. Brook Taylor, a disciple of Newton, stated the general result in his *Methodus Incrementorum Directa et Inversa* published in 1715. Taylor based his derivation on the interpolation formula due to Isaac Newton and the Scottish mathematician James Gregory. Although it is not clear that Gregory had the general formula in hand, it appears that he could derive the power series for any particular function as early as 1671, 44 years before Taylor. Later, Joseph-Louis de Lagrange gave Taylor series a central role in his treatment of calculus but mistakenly assumed that any continuous function can be expanded in a Taylor series. Historically, Taylor series paved the way for the study of infinite series expansions of functions. Equally important to operations research, Taylor series inaugurated approximation theory by using a polynomial function to approximate a suitably differentiable function with a known error bound. [*The Historical Development of the Calculus*, C. H. Edwards, Jr., Springer-Verlag, New York, 1979; *Mathematics and its History*, J. Stillwell, Springer-Verlag, New York, 1989]

1718 *The Doctrine of Chances*, Abraham de Moivre

The three editions of this classic book of Abraham de Moivre defined the course of probability theory from 1718 to 1756. The book consists of an introduction with elementary probability theorems, followed by a collection of problems. The first edition contains 53 problems on probability, while the second edition of 1738 has 75 problems on probability and 15 on insurance mathematics. Due to his advanced age and failing eyesight, de Moivre was forced to entrust the last revision to a friend. The last edition of 1756 was published posthumously and includes 74 problems on probability and 33 on insurance mathematics.

The importance of this text was recognized by both Joseph-Louis de Lagrange and Pierre-Simon Laplace, who independently planned to translate it. [*Games, Gods, and Gambling: A History of Probability and Statistical Ideas*, F. N. David, C. Griffin, London, 1962 (Dover reprint 1998); *A History of Probability and Statistics and Their Applications Before 1750*, A. Hald, John Wiley & Sons, New York, 1990]

De Moivre and Newton at Starbucks:

De Moivre studied mathematics at the Sorbonne before emigrating to England in 1688, where he earned a living as tutor to the sons of several noblemen. According to David and Griffin (1962), de Moivre came across a copy of Newton's *Principia Mathematica* at the house of one of his students. As he found the subject matter beyond him, he obtained a copy, tore it into pages, and so learned it "page by page as he walked London from one tutoring job to another." Later, de Moivre became friends with Newton and they would meet occasionally in de Moivre's favorite coffee shop. They often went to Newton's home to continue their conversation. When Newton became Master of the Mint (1703), his interest in mathematical exposition waned. When approached by students, Newton would say: "Go to Mr. de Moivre; he knows these things better than I do."

Abraham de Moivre
1667–1754

1733 First appearance of the normal distribution

Abraham de Moivre stated a form of the central limit theorem (the mean of a random sample from any distribution is approximately distributed as a normal variate) by establishing the normal approximation to the binomial distribution. De Moivre derived this result when he was 66 years of age and incorporated it into the second edition of his book, *Doctrine of Chances* (1738). Other mathematicians, Karl Friedrich Gauss, Joseph-Louis de Lagrange and Pierre-Simon Laplace, were influenced by de Moivre's work, with Gauss rediscovering the normal curve in 1809, and Laplace in 1812 with his publication of *Théorie analytique des probabilités*. ["Abraham De Moivre's 1733 derivation of the normal curve: A bibliographic note," R. H. Daw, E. S. Pearson, *Biometrika*, 59, 1972, 677–680; *The History of Statistics*, S. M. Stigler, Harvard University Press, Cambridge, 1986; *Mathematical Methods of Statistics*, H. Cramér, Harvard University Press, Cambridge, 1946]

1733 Beginnings of geometric probability

George-Louis Leclerc, Comte de Buffon, had broad interests in natural history, mathematics, and statistics. Wishing to demonstrate that "chance falls within the domain of geometry as well as analysis," Buffon presented a paper on the game of *franc-carreau* in which he analyzed a problem in geometrical probability. This paper makes mention of the famous eponymous needle problem. Buffon is considered to be a precursor of demographics due to his use of real data in analyzing the statistics of mortality and life expectancy. [*Statisticians of the Centuries*, G. C. Heyde, E. Seneta, editors, Springer-Verlag, New York, 2001]

Drop the needle:

Buffon's famous needle problem can be used to experimentally determine an approximate value of π: Rule a large plane area with equidistant parallel straight lines. Throw (drop) a thin needle at random on the plane. Buffon showed that the probability that the needle will fall across one of the lines is $2l/\pi d$, where d is the distance between the lines and l is the length of the needle, with $l < d$.

Comte de Buffon
1707–1788

1736 Königsberg bridge problem

Leonhard Euler, a Swiss mathematician, is credited with establishing the theory of graphs. His relevant paper described the city of Königsberg's seven bridge configuration that joined the two banks of the Pregel River and two of its islands, and answered the question: Is it possible to cross the seven bridges in a continuous walk without recrossing any of them? The answer was no. Euler showed that for such a configuration (graph) to have such a path, the land areas (nodes) must be connected with an even number of bridges (arcs) at each node. ["Solutio Problematis Ad Geometriam Situs Pertinentis," L. Euler, *Commentarii Academiae Scientiarum Imperialis Petropolitanae*, 8, 1736, 128–140 (translated in *Graph Theory 1736–1936*, N. L. Biggs, E. K. Lloyd, R. J. Wilson, Oxford University Press, Oxford, 1976, 157–190); *Graphs and Their Uses*, O. Ore, Random House, New York, 1963; *Combinatorial Optimization: Networks and Matroids*, E. Lawler, Holt, Rinehart and Winston, New York, 1976; *Graphs as Mathematical Models*, G. Chartrand, Prindle, Weber & Schmidt, Boston, 1977]

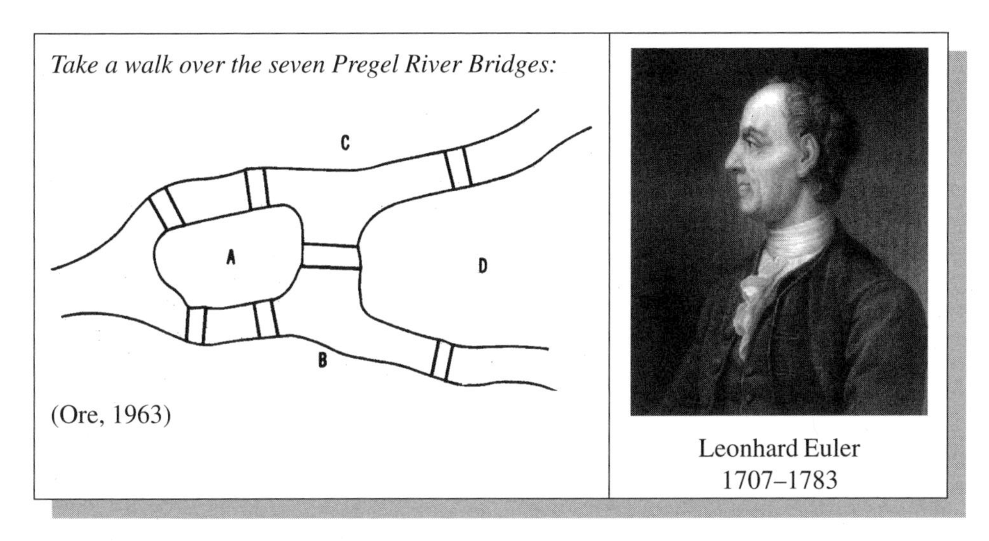

Leonhard Euler
1707–1783

1755 Least absolute deviation (L_1) regression

Rogerius Josephus Boscovich, a mathematics professor at the *Collegium Romanum* in Rome, developed the first objective procedure for fitting a linear relationship to a set of observations. He posed the problem of finding the values of coefficients a and b to fit n equations of the form $y_i = a + bx_i$ $(i = 1, \ldots, n)$. Initially, Boscovitch considered taking the average of the individual slopes $(y_i - y_j)/(x_i - x_j)$ computed for all pairs (i, j) with $i < j$, but eventually settled on the principle that a and b should be chosen to ensure an algebraic sum of zero for the error terms $e_i = y_i - (a + bx_i)$, and to minimize the sum of the absolute values of these terms. An efficient algorithm for finding the regression coefficients for the general case had to await linear programming. ["R. J. Boscovich's work on probability," O. B. Sheynin, *Archive for History of Exact Sciences*, 9, 1973, 306–32; *Statisticians of the Centuries*, G. C. Heyde, E. Seneta, editors, Springer-Verlag, New York, 2001]

1763 Bayes Rule

The Reverend Thomas Bayes proposed a rule (formula) for estimating a probability p by combining *a priori* knowledge of p with information contained in a finite number of n current independent trials. Let the collection of events $\{H_i\}$ be n mutually exclusive and exhaustive events. Let E be an event for which we know the conditional probabilities $P[E|H_i]$ of E, given H_i, and also the absolute *a priori* probabilities $P[H_i]$. Then Bayes rule enables us to determine the conditional *a posteriori* probability $P[H_i|E]$ of any of the events H_i. If the events $\{H_i\}$ are considered as "causes," then Bayes rule can be interpreted as a formula for the probability that the event E, which has occurred, is the result of "cause" H_i. Bayes rule forms the basis of the subjective interpretation of probability. ["An essay towards solving a problem in the doctrine of chances," T. Bayes, *Philosophical Transactions of the Royal Society of London*, 53, 1763, 370–418 (reprinted in *Biometrika*, 45, 1958, 293–315); *An Introduction to Probability Theory and its Applications*, W. Feller, John Wiley & Sons, New York, 1950; *Modern Probability Theory and its Applications*, E. Parzen, John Wiley & Sons, New York, 1960]

Bayes Rule:

$$P[H_i|E] = \frac{P[EH_i]}{P[E]} = \frac{P[E|H_i]P[H_i]}{\sum_{j=1}^{n} P[E|H_j]P[H_j]}$$

Thomas Bayes
1702–1761

1788 Lagrange multipliers

The French mathematician Joseph-Louis de Lagrange's celebrated book, *Mécanique Analytique*, included his powerful method for finding extrema of functions subject to equality constraints. It was described here as a tool for finding the equilibrium state of a mechanical system. If $f(x)$ denotes the potential function, the problem is to minimize $f(x)$ subject to $g_i(x) = 0$ for $i = 1, \ldots, m$. The Lagrangian necessary condition for equilibrium states that at the minimizing point x^*, the gradient of $f(x)$ can be expressed as a linear combination of the gradients of the $g_i(x)$. The factors that form the linear combination of these gradients are called Lagrange multipliers. The important case of inequality constraints was first investigated by the French mathematician Jean-Baptiste-Joseph Fourier: Minimize $f(x)$ subject to $g_i(x) \geqslant 0$ for $i = 1, \ldots, m$. The comparable necessary condition states that the gradient of $f(x)$ can be expressed as a *nonnegative* linear combination of the gradients of the $g_i(x)$. This condition was stated without proof by the French economist-mathematician Antoine-Augustin Cournot (1827) for special cases, and by the Russian mathematician Mikhail Ostrogradski (1834) for the general case. The Hungarian mathematician Julius (Gyula) Farkas supplied the first complete proof in 1898. ["Generalized Lagrange multiplier method for solving problems of optimum allocation of resources," H. Everett, III, *Operations Research*, 11, 1963, 399–417; "On the development of optimization theory," A. Prékopa, *American Mathematical Monthly*, 87, 1980, 527–542]

Joseph-Louis de Lagrange
1736–1813

1789 Principle of utility

Jeremy Bentham, an English jurist and philosopher, published *An Introduction to the Principles of Morals and Legislation* in which he proclaimed that mankind is governed by pain and pleasure, and proposed a principle of utility "... which approves or disapproves of every action whatsoever, according to the tendency which it appears to have to augment or diminish the happiness of the party whose interest is in question." Or, in general, that the object of all conduct or legislation is "the greatest happiness for the greatest number." Bentham's writings are considered to be the precursors of modern utility theory. ["An introduction to the principles of morals and legislation," J. Bentham, 1823, pp. 3–29 in *Utility Theory: A Book of Readings*, A. N. Page, editor, John Wiley & Sons, New York, 1968; *Works of Jeremy Bentham*, J. Bentham, Tait, Edinburgh, 1843; *Webster's New Biographical Dictionary*, Merriam-Webster, Springfield, 1988]

Bentham's Felicific Calculus:

For a particular action, Bentham suggests measuring pleasure or pain using six dimensions of value (criteria): its intensity, its duration, its certainty or uncertainty, its propinquity or remoteness (nearness in time or place), its fecundity (chance of being followed by sensations of the same kind), its purity (chance of not being followed by sensations of the opposite kind). The indivdual or group contemplating the action then sums up all the delineated pleasures and pains and takes the balance; one adds positive pleasure values to negative pain values to obtain a final happiness score for the action. Bentham's Felicific Calculus leads directly to the modern basic problem of decision analysis: How to select between alternatives or how to rank order alternatives?

At University College, London, a wooden cabinet contains Bentham's preserved skeleton, dressed in his own clothes, and surmounted by a wax head. Bentham had requested that his body be preserved in this way.

Jeremy Bentham
1748–1832

1795 Method of least squares

The German mathematician Carl Friedrich Gauss and French mathematician Adrien-Marie Legendre are both credited with independent discovery of the method of least squares, with Gauss' work dating from 1795 and Legendre publishing his results, without proof, in 1805. The first proof that the method is a consequence of the Gaussian law of error was published by Gauss in 1809. Robert Adrian, an Irish mathematician who emigrated to the U.S., unaware of the work of Gauss and Legendre, also developed and used least squares, circa 1806. Least squares, so named by Legendre, is the basic method for computing the unknown parameters in the general regression model which arises often in applications of operations research and related statistical analyses. [*A History of Mathematics*, C. B. Boyer, John Wiley & Sons, New York, 1968; *Encyclopedia of Statistical Sciences*, Vol. 4, S. Kotz, N. L. Johnson, editors, John Wiley & Sons, New York, 1982; *Applied Linear Statistical Models*, 3rd edition, J. Neter, W. Waserman, M. K. Kutner, Irwin, Homewood, 1990]

Carl Friedrich Gauss
1777–1855

Adrien-Marie Legendre
1752–1833

1810 The general central limit theorem

Pierre-Simon Laplace derived the general central limit theorem: The sum of a sufficiently large number of independent random variables follows an approximately normal distribution. His work brought an unprecedented new level of analytical techniques to bear on probability theory. [*The History of Statistics*, S. M. Stigler, Harvard University Press, Cambridge, 1986; *Pierre-Simon Laplace 1749–1827: A Life in Exact Science*,

C. C. Gillispie, Princeton University Press, Princeton, 1997; *Statisticians of the Centuries*, G. C. Heyde, E. Seneta, editors, Springer-Verlag, New York, 2001, 95–100]

"... all our knowledge is problematical":

Laplace's book, *Théorie analytique des probabilities* first appeared in 1812 and remained the most influential book on mathematical probability to the end of the nineteenth century. Aiming at the general reader, Laplace wrote an introductory essay for the second (1814) edition. This essay, *A Philosophical Essay on Probabilities*, explained the fundamentals of probability without using higher mathematics. It opened with: *"... all our knowledge is problematical, and in the small number of things which we are able to know with certainty, even in the mathematical sciences themselves, the principal means of ascertaining truth – induction and analogy – are based on probabilities; so that the entire system of human knowledge is connected with the theory set forth in this essay."*

Pierre-Simon Laplace
1749–1827

1811 Kriegsspiel (war gaming)

A rule-based (rigid) process based on actual military operations that uses a map, movable pieces that represent troops, two players and an umpire was invented by the Prussian War Counselor von Reisswitz and his son, a lieutenant in the Prussian army. It was modified in 1876 by Colonel von Verdy du Vernois into free kriegspiel that imposed simplified rules and allowed tactical freedom. [*Fundamentals of War Gaming*, 3rd edition, F. J. McHugh, U.S. Naval War College, Newport, 1966; "Military Gaming," C. J. Clayton, pp. 421–463 in *Progress in Operations Research*, Vol. I, R. L. Ackoff, editor, John Wiley & Sons, New York, 1961; *The War Game*, G. D. Brewer, M. Shubik, Harvard University Press, Cambridge, 1979]

1826 Solution of inequalities

Jean-Baptiste-Joseph Fourier, a French mathematician, is credited with being the first one to formally state a problem that can be interpreted as being a linear-programming problem. It dealt with the solution of a set of linear inequalities. ["Solution d'une question particulière du calcul des inégalités," J. Fourier, *Nouveau Bulletin des Sciences par la Société philomathique de Paris*, 1826, 99–100; "Joseph Fourier's anticipation of linear programming," I. Grattan-Guiness, *Operational Research Quarterly*, 21, 3, 1970, 361–364]

Jean-Baptiste-Joseph Fourier
1768–1830

1826 Solution of linear equations

Carl Friedrich Gauss used elementary row operations (elimination) to transform a square ($n \times n$) matrix A, associated with a set of linear equations, into an upper triangular matrix U. Once this is accomplished, it is a simple matter to solve for variable x_n and then, by successive back-substitution, to solve for the other variables by additions and subtractions. This process has been modified to the Gauss–Jordan elimination method in which A is transformed into a diagonal matrix D that allows the values of the variables to computed without any back substitutions. ["Theoria Combinationis Observationum Erroribus Minimis Obnoxiae," C. F. Gauss, *Werke*, Vol. 4, Göttingen, 1826; *A Handbook of Numerical Matrix Inversion and Solution of Linear Equations*, J. R. Westlake, Krieger Publishing, New York, 1975]

1833 Analytical Engine

Charles Babbage, an English mathematician and inventor, is credited with being the first one to conceive a general purpose computer (Analytical Engine). Although never built *in toto*, its design employed punch cards for data and for defining a set of instructions (program). Powered by steam, it would have been able to store a thousand fifty-digit numbers. [*The Computer from Pascal to von Neumann*, H. M. Goldstine, Princeton University Press, Princeton, 1972; *A Computer Perspective*, G. Fleck, editor, Harvard University Press, Cambridge, 1973; *Webster's New Biographical Dictionary*, Merriam-Webster, Springfield, 1988; *The Difference Engine: Charles Babbage and the Quest to Build the First Computer*, D. Swade, Viking/Penguin-Putnam, New York, 2000]

On mail and cows:

Babbage is considered to be an early operations researcher (the first?) based on his on-site analysis of mail handling costs in the British Post Office (see his book *On the Economy of Machinery and Manufacturers*, 1832). He also invented the locomotive cowcatcher.

Charles Babbage
1792–1871

1837 The Poisson approximation

Sequences of independent Bernoulli trials, where each trial has only two outcomes, success with a probability of p and failure with a probability of $(1 - p)$, were studied by Jakob Bernoulli I, Abraham de Moivre and a number of other mathematicians. The French mathematician Siméon-Denis Poisson was known for his "law of large numbers" that counted the proportion of successes in such sequences when the probability p could vary from one trial to the next. Today, Poisson's name is more readily associated with his approximation for the binomial distribution which counts the number of successes in n independent Bernoulli trials with the same p. Poisson first expressed the cumulative terms of the binomial distribution in terms of the negative binomial distribution and then considered the limit as n goes to infinity and p goes to zero in such a way that $\lambda = np$ remains fixed. The approximation resulted in cumulative terms of the Poisson probability mass function $P(k) = \mathrm{e}^{-\lambda}(\lambda^k/k!)$, the probability of k successes. Curiously, the Poisson probability law or any distribution of that form is not explicitly found in Poisson's writings. [*An Introduction to Probability Theory and its Application*, W. Feller, John Wiley & Sons, New York, 1950; "Poisson on the Poisson distribution," S. M. Stigler, *Statistics and Probability Letters*, 1, 1982, 33–35; *A History of Probability and Statistics and Their Applications Before 1750*, A. Hald, John Wiley & Sons, New York, 1990; "The theory of probability," B. V. Gnedenko, O. B. Sheinin, Chapter 4 in *Mathematics of the 19th Century*, A. N. Kolmogorov, A. P. Yushkevich, editors, Birkäuser Verlag, Boston, 2001]

Siméon-Denis Poisson
1781–1840

1839 Founding of the American Statistical Society

The American Statistical Society (ASA) was founded in Boston in 1839, making it the second oldest professional society in the United States. ASA's mission is to promote statistical practice, applications, and research; publish statistical journals; improve statistical education; and advance the statistics profession. Its first president was Richard Fletcher. [www.amstat.org]

Early statisticians of note:

Members of the ASA included President Martin Van Buren, Florence Nightingale, Andrew Carnegie, Herman Hollerith, and Alexander Graham Bell.

1845 Network flow equations

The German physicist Gustav R. Kirchhoff discovered two famous laws that describe the flow of electricity through a network of wires. Kirchhoff's laws, the conservation of flow at a node (in an electrical circuit, the currents entering a junction must equal the currents leaving the junction), and the potential law (around any closed path in an electrical circuit the algebraic sum of the potential differences equals zero) have a direct application to modern networks and graphs. Kirchhoff also showed how to construct a fundamental set of $(n + m - 1)$ circuits in a connected graph with m nodes and n edges. [*Graph Theory 1736–1936*, N. L. Biggs, E. K. Lloyd, R. J. Wilson, Oxford University Press, Oxford, 1976; *Network Flow Programming*, P. A. Jensen, J. W. Barnes, John Wiley & Sons, New York, 1980; *Webster's New Biographical Dictionary*, Merriam-Webster, Springfield, 1988]

1846 Fitting distributions to social populations

In his book, *Sur l'homme et le développement de ses facultés* (1835), the Belgian statistician Adolphe Quetelet presented his ideas on the application of probability theory to the study of human populations and his concept of the average man. Quetelet also pioneered the fitting of distributions to social data. In this effort, he was struck by the widespread occurrence of the normal distribution. His approach to the fitting of normal curves is explained in letters 19–21 of his 1846 book, a treatise written as a collection of letters to the Belgian King's two nephews, whom Quetelet had tutored. One of the data sets on which Quetelet demonstrated his fitting procedure is among the most famous of the nineteenth century, the frequency distribution of the chest measurements of 5732 Scottish soldiers. [*Lettres à S. A. R. Le Duc Régnant de Saxe-Cobourget et Gotha sur la Théorie des Probabilités appliqués aux sciences morale et politiques*, A. Quetelet, Hayez, Brussels, 1846; *The History of Statistics*, S. M. Stigler, Harvard University Press, Cambridge,

1986; *The Politics of Large Numbers*, A. Desrosières, Harvard University Press, Cambridge, 1998]

1856 Hamiltonian cycles

Given a graph of edges and vertices, a closed path that visits all vertices of a graph exactly once is called a Hamiltonian cycle. How to find such a cycle is an important problem in network analysis. Early versions of this problem considered finding a knight's tour (a Hamiltonian cycle for all 64 squares on a chessboard). The cycle is named after the Irish mathematician Sir William R. Hamilton. [*Graph Theory 1736–1936*, N. L. Biggs, E. K. Lloyd, R. J. Wilson, Oxford University Press, Oxford, 1976; *The Traveling Salesman Problem: A Guided Tour of Combinatorial Optimization*, E. L. Lawler, J. K. Lenstra, A. H. G. Rinnooy Kan, D. B. Shmoys, editors, John Wiley & Sons, New York, 1985]

Cycling with Hamilton:

Hamilton created a game called the Icosian Game that requires the finding of Hamiltonian cycles through the 20 vertices that are connected by the 30 edges of a regular solid dodecahedron.

William R. Hamilton
1805–1865

1873 Solution of equations in nonnegative variables

The importance of nonnegative solutions to sets of inequalities and equations was not evident until the development of linear programming. Earlier work, that comes under the modern heading of transposition theorems, is illustrated by the German mathematician P. Gordan's theorem: There is a vector x with $x \geqslant 0$, $x \neq 0$, if and only if there is no vector y with $yA > 0$. ["Über die Auflösung linearer Gleichungen mit reellen Coefficienten," P. Gordan, *Mathematische Annalen*, 6, 1873, 23–28; *Theory of Linear and Integer Programming*, A. Schrijver, John Wiley & Sons, New York, 1986]

1873 Galton's quincunx

The English statistician Francis Galton designed the quincunx to illustrate how the normal distribution could arise due to random events. The name stems from an arrangement of five objects, one at each corner of a rectangle or square and one at the center. Galton's quincunx consisted of a glass encased vertical board with a succession of offset rows of equally spaced pins top to bottom. Each nail is directly below the midpoint of two adjacent nails in the row above. Thus, except for those at the boundary, each nail is the center of a square quincunx of five nails. A funnel at the top allows lead shot to fall down while bouncing against the pins, resulting in a random walk with a 50–50 chance of going left or right. The shots are collected in a set of compartments as they fall to the bottom. This ping-ponging of the shot against the pins yields frequency counts in the compartments in the form of a binomial histogram ($p = 1/2$) that produces a visual approximation of the normal distribution. The quincunx illustrates how a large number of random accidents give rise to the normal distribution. Galton described it as an "instrument to illustrate the law of error or dispersion." Karl Pearson constructed a quincunx in which the value of p can be varied, thus producing skewed binomial distributions. ["Quincunx", H. O. Posten, pp. 489–491 in *Encyclopedia of Statistical Sciences*, Vol. 7, S. Kotz, N. L. Johnson, editors, John Wiley & Sons, New York, 1982; *The History of Statistics*, S. M. Stigler, Harvard University Press, Cambridge, 1986; *Statistics on the Table*, S. M. Stigler, Harvard University Press, Cambridge, 1999]

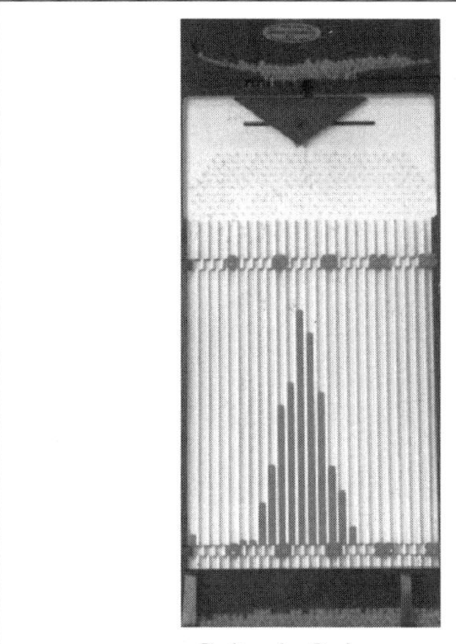

Galton's Quincunx

Francis Galton
1822–1911

2

Operations research precursors from 1881 to 1935

1881 Scientific management/Time studies

Frederick W. Taylor, an American engineer and management consultant, is called "the father of Scientific Management." Taylor introduced his seminal time study method in 1881 while working as a general plant foreman for the Midvale Steel Company. He was interested in determining answers to the interlocking questions of "Which is the best way to do a job?" and "What should constitute a day's work?" As a consultant, he applied his scientific management principles to a diverse set of industries. [*The Principles of Scientific Management*, F. W. Taylor, Harper & Brothers, New York, 1911; *Motion and Time Study: Design and Measurement of Work*, 6th edition, R. M. Barnes, John Wiley & Sons, New York, 1968; *Executive Decisions and Operations Research*, D. W. Miller, M. K. Starr, Prentice-Hall, Englewood Cliffs, 1969; *Work Study*, J. A. Larkin, McGraw-Hill, New York, 1969; *A Computer Perspective*, G. Fleck, editor, Harvard University Press, Cambridge, 1973; *Webster's New Biographical Dictionary*, Merriam-Webster, Springfield, 1988; *The One Best Way: Frederick Winslow Taylor and the Enigma of Efficiency*, R. Kanigel, Viking, New York, 1997]

Early Operations Research:

A definition of Taylorism could be confused with an early definition of OR as it moved away from its military origins: "The application of scientific methods to the problem of obtaining maximum efficiency in industrial work or the like," Kanigel (1997).

Taylor, while working for Bethlehem Steel Company (1898), concluded, by observation and experimentation, that to maximize a day's workload when shoveling ore, a steel-mill workman's shovel should hold 21½ pounds. Taylor's motto: "A big day's work for a big day's pay."

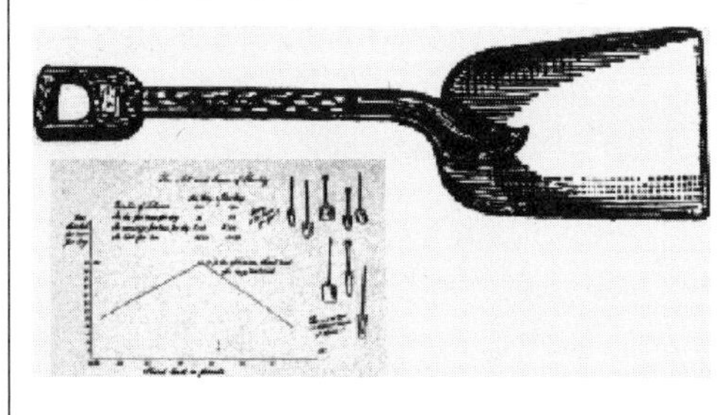

Frederick W. Taylor
1856–1915

1885 Scientific management/Motion studies

More or less coincident with Frederick W. Taylor's time studies was the development of motion studies by Frank B. Gilbreth. In his first job for a building contractor (in 1885), Frank Gilbreth, at the age of 17, made his first motion study with the laying of bricks. He later formed a consulting engineering firm with his wife, Lilllian M. Gilbreth. They were concerned with "eliminating wastefulness resulting from using ill-directed and inefficient motions." As noted by Larkin (1969): "Time and motion study originates from a marriage of Gilbreth's motion study with what was best in Taylor's investigational techniques." The Gilbreths, Taylor and Henry L. Gantt, who worked with Taylor, are considered to be the pioneers of scientific management. [*Motion Study*, F. Gilbreth, D. Van Nostrand Co., New York, 1911; *Cheaper by the Dozen*, F. B. Gilbreth, Jr., E. Gilbreth Carey, Thomas Y. Crowell Company, New York, 1949; *Motion and Time Study: Design and Measurement of Work*, 6th edition, R. M. Barnes, John Wiley & Sons, New York, 1968; *Executive Decisions and Operations Research*, D. W. Miller, M. K. Starr, Prentice-Hall, Englewood Cliffs, 1969; *The Frank Gilbreth Centennial*, The American Society of Mechanical Engineers, New York, 1969; *Work Study*, J. A. Larkin, McGraw-Hill, New York, 1969]

Bricks and baseball:

In his brick laying motion study, Frank Gilbreth invented an adjustable scaffold and reduced the motions per brick from 18 to 5, with the bricklaying rate increasing from 120 to 350 per hour.

Gilbreth made a film of the Giants and the Phillies baseball game, Polo Grounds, May 31, 1913. He determined that a runner on first, who was intent on stealing second base and had an eight foot lead, would have to run at a speed faster than the world's record for the 100-yard dash.

Frank Gilbreth
1868–1924

The first lady of engineering:

Lillian Gilbreth teamed with her husband to conduct a number of motion studies and to write many books describing their methodology. She was an engineer and a professor of management at Purdue University and the University of Wisconsin. She was also the mother of 12 children. The exploits of the Gilbreth family and their children were captured in the book *Cheaper by the Dozen* and in the 1950 movie starring Clifton Webb and Myrna Loy.

Lillian Gilbreth
1878–1972

1890 Statistical simulation with dice

Francis Galton described how three dice can be employed to generate random error terms that corresponded to a discrete version of half-normal variate with median error of 1.0. By writing four values along the edges of each face of the die, Galton could randomly generate 24 possibilities with the first die, use a second die to refine the scale, and a third to identify the sign of the error. Providing an illustration of these dice, Stigler calls them "perhaps the oldest surviving device for simulating normally distributed random numbers." Earlier, Erastus Lyman de Forest had used labeled cards and George H. Darwin relied on a spinner to generate half normal variates. Galton states that he had a more general approach in mind. ["Stochastic Simulation in the Nineteenth Century," *Statistics on the Table*, S. M. Stigler, Harvard University Press, Cambridge, 1999]

1896 Geometry of numbers

The Russian-born, German mathematician Hermann Minkowski is considered the father of convex analysis. In his pathbreaking treatise on the geometry of numbers, Minkowski used the tools of convexity to approach number theory from a geometrical point of view. One fundamental question was to identify conditions under which a given region contains a lattice point – a point with integer coordinates. In the case of the plane, Minkowski's fundamental theorem states that any convex set that is symmetric about the origin and has area greater than 4 contains non-zero lattice points. Minkowski's work has important implications for the diophantine approximations (using rationals of low denominator to approximate real numbers) and systems of linear inequalities in integer variables. More than 80 years later, Hendrick W. Lenstra, Jr. introduced methods from the geometry of numbers into integer programming using an efficient algorithm for basis reduction. [*Geometrie der Zahlen*, H. Minkowski, Teubner, Leipzig, 1896; "Integer programming with a fixed number of variables," H. W. Lenstra, Jr., *Mathematics of Operations Research*, 8, 1983, 538–548; *Geometric Algorithms and Combinatorial Optimization*, M. Grötschel, L. Lovász, A. Shrijver, Springer-Verlag, New York, 1988; *The Geometry of Numbers*, C. D. Olds, A. Lax, G. Davidoff, The Mathematical Association of America, Washington, DC, 2000]

1896 Representation of convex polyhedra

A polyhedral convex set is defined by $P = \{x \in R^n \mid Ax \leqslant b\}$. The Representation Theorem states that any point of P can be represented as a convex combination of its extreme points plus a non-negative combination of its extreme directions (i.e., finitely generated). This result is central to linear programming and the computational aspects of the simplex method. Hermann Minkowski first obtained this result for the convex cone $\{x \in R^n \mid Ax \leqslant 0\}$ (Schrijver, 1986). Minkowski's result was also known to Julius Farkas and was refined by Constantin Carathéodory. The general statement of the Representation Theorem – a convex set is polyhedral if and only if it is finitely generated – is due to Hermann Weyl (1935). Rockafellar comments: "This classical result is an outstanding example of a fact that is completely obvious to geometric intuition, but which wields important algebraic content and is not trivial to prove." An equivalent result is Theodore Motzkin's Decomposition Theorem: any convex polyhedron is the sum of a polytope and a polyhedral cone. [*Geometrie der Zahlen*, H. Minkowski, Teubner, Leipzig, 1896; "Uber den Variabilitatsbereich der Koeffizienten von Potenzreihen, die gegebene Werte nicht annehmen," C. Carathéodory, *Mathematische Annalen*, 64, 1907, 95–115; "Elemantere Theorie der konvexen polyeder," H. Weyl, *Commentarii Math. Helvetici*, 7, 1935, 290–235; *Beiträge zur Theorie der Linearen Ungleichungen*, T. Motzkin, Doctoral Thesis, University of Zurich, 1936; *Convex Analysis*, R. Tyrell Rockafellar, Princeton University Press, Princeton, 1963; *Theory of Linear and Integer Programming*, A. Shrijver, John Wiley & Sons, New York, 1986; *Linear Optimization and Extensions*, 2nd edition, M. Padberg, Springer-Verlag, New York, 1999]

Space-time connections:

Hermann Minkowski was raised in Königsberg where he and David Hilbert were fellow university students. They later became colleagues at Göttingen. Hermann Weyl completed his doctorate with Hilbert, while Carathéodory worked on his with Minkowski. Both Minkoswki and Weyl are known for their contributions to mathematical physics and the geometry of space-time. Minkowski's research on the geometry of space-time was motivated by his close reading of the 1905 paper on special relativity by Albert Einstein, his former student. (Padberg, 1999).

Hermann Minkowski
1864–1909

1900 Gantt charts

Henry L. Gantt, an associate of Frederick Taylor, devised a project planning method by which managers could depict, by a sequence of bars on a chart, a project's interrelated steps, show precedence relationships between steps, indicate completion schedules, and track actual performance. It is still a basic management tool, especially in the construction industry. [*Executive Decisions and Operations Research*, D. W. Miller, M. K. Starr, Prentice-Hall, Englewood Cliffs, 1969; *Introduction to Operations Research*, 7th edition, F. S. Hiller, G. J. Lieberman, McGraw-Hill, New York, 2001; *The Informed Student Guide to Management Science*, H. G. Daellenbach, R. L. Flood, Thompson, London, 2002]

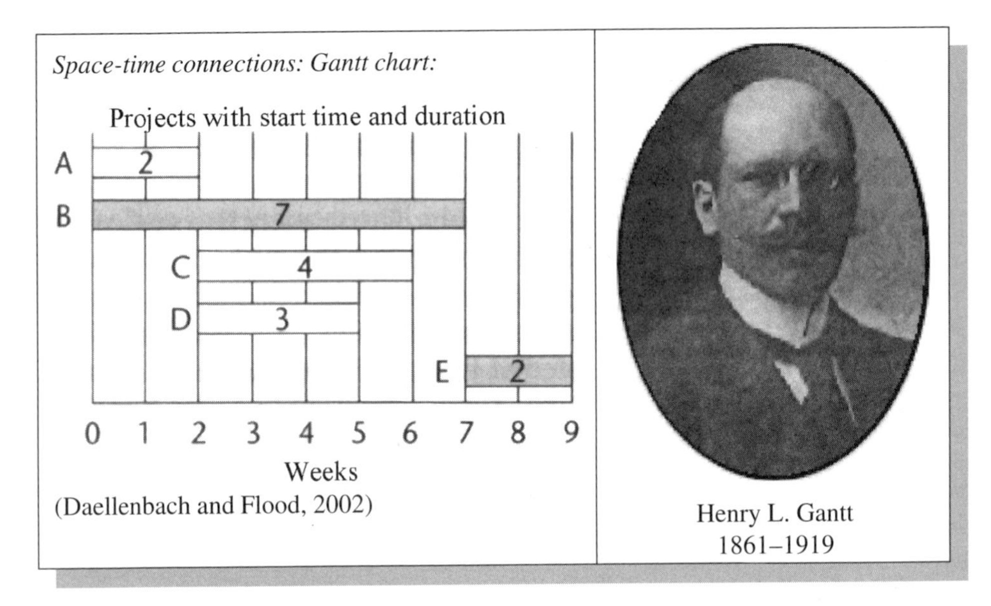

Henry L. Gantt
1861–1919

1900 Brownian motion applied to the stock market

A student of Henri Poincaré, Louis Bachelier, in his doctoral thesis, *Théorie de la spéculation*, proposed the application of "the calculus of probabilities to stock market operations." This work contains the first treatment of Brownian motion to stock markets, providing three different characterizations. The results, although essentially correct, were unjustly regarded as imprecise or vague and did not receive due recognition. Bachelier also considered what is now termed the drift of a stochastic differential equation. The full recognition of his work had to wait till the 1970s, when the theory of options trading gained currency. [*Statisticians of the Centuries*, G. C. Heyde, E. Seneta, editors, Springer-Verlag, New York, 2001]

1900 Early result on total unimodularity

Matrix A is totally unimodular if each subdeterminant of A has a value of 0, 1, or -1. Henri Poincaré was the first to state that a matrix A with all entries a_{ij} equal to 0, $+1$, or -1 is totally unimodular if A has exactly one $+1$ and exactly one -1 in each column (and zeros otherwise). Poincaré derived this fact from a more general result involving cycles composed on entries of A. Much later, Alan Hoffman and Joseph B. Kruskal showed that unimodularity was the fundamental reason why certain classes of linear programs have integral optimal solutions. ["Integral boundary points of convex polyhedra," A. J. Hoffman, J. B. Kruskal, pp. 223–246 in *Linear Inequalities and Related Systems*, H. W. Kuhn, A. W. Tucker, editors, Princeton University Press, Princeton, 1956; *Theory of Linear and Integer Programming*, A. Schrijver, John Wiley & Sons, New York, 1986, pp. 266–279, 378]

1900 Chi-Square Test

Karl Pearson
1857–1936

At the turn of the century, the British statistician Karl Pearson devised the chi-square goodness of fit test, a fundamental advance in the development of the modern theory of statistics. The test determines the extent of the fit of a set of observed frequencies of an empirical distribution with expected frequencies of a theoretical distribution. ["Karl Pearson and degrees of freedom," pp. 338–357 in *Statistics on the Table*, Stephen M. Stigler, Harvard University Press, Cambridge, 1999; *Statisticians of the Centuries*, G. C. Heyde, E. Seneta, editors, Springer-Verlag, New York, 2001]

1901 Solution of inequality systems

The duality theorem of linear programming that relates the solution to the primal and dual problems was first proved by David Gale, Harold W. Kuhn and Albert W. Tucker in 1951 using the 1902 theorem of the Hungarian mathematician Julius (Gyula) Farkas. Given the set of homogeneous inequalties (1) $g_i^{\mathrm{T}} x \geqslant 0$ $(i = 1, \ldots, m)$ and (2) $g^{\mathrm{T}} x \geqslant 0$, where the g_i, g and x are n-component vectors. The inequality (2) is a consequence of the inequalities (1) if and only if there are nonnegative numbers $\lambda_1, \ldots, \lambda_m$ such that $g = \lambda_1 g_1 + \cdots + \lambda_m g_m$. ["Über die Theorie der einfachen Ungleichungen," J. Farkas, *J. Reine Angew. Math.*, 124, 1901(2), 1–24; "Linear programming and the theory of games," D. Gale, H. W. Kuhn, A. W. Tucker, pp. 317–329 in *Activity Analysis of Production and Allocation*, T. C. Koopmans, editor, John Wiley & Sons, New York, 1951; "On the development of optimization theory," A. Prékopa, *American Mathematical Monthly*, 1980, 527–542]

1906 Pareto optimal solution

Vilfredo Pareto
1848–1923

The Italian economist Vilfredo Pareto proposed that in competitive situations a solution is optimum (efficient) if no actor's satisfaction can be improved without lowering (degrading) at least one other actor's satisfaction level. That is, you cannot rob Peter to pay Paul. In multi-objective situations, a Pareto optimum is a feasible solution for which an increase in value of one objective can be achieved only at the expense of a decrease in value of at least one other objective. [*Manuale di econnomia politica*, V. Pareto, Società Editrice Libraria, Milano, 1906; *Three Essays on the State of Economic Science*, T. C. Koopmans, McGraw-Hill, New York, 1957; *Theory of Value*, G. Debreu, John Wiley & Sons, New York, 1959]

1907 Markov process and chain

The Russian mathematician Andrei Andreevich Markov developed the concept of a Markov process from his studies on sequences of experiments "connected in a chain." A Markov process has the property that, given the value of the time dependent random variable X_t, the values of X_s, $s > t$, do not depend on the values of X_u, $u < t$. This is known as the lack-of-memory property for such processes: the probabilities of future events are completely determined by the present state of the process and the probabilities of its behavior from the present state on. The research and writings of the statistician William W. Feller brought Markov processes and chains to the attention of the operations research and statistical communities. ["Investigation of a noteworthy case of dependent trials," A. A. Markov, *Izv. Ros. Akad. Nauk*, 1, 1907; *An Introduction to Probability Theory and Its Applications*, W. W. Feller, John Wiley & Sons, New York, 1950; *Markov Processes*, Vols. I and II, E. B. Dynkin, Academic Press, New York, 1965; *A First Course in Stochastic Processes*, S. Karlin, Academic Press, New York, 1966; *Encyclopedia of Operations Research and Management Science*, 2nd edition, S. I. Gass, C. M. Harris, editors, Kluwer Academic Publishers, Boston, 2001]

Andrei A. Markov
1856–1922

1908 Student's t-distribution

Better known by his *nom de plume* "Student," William Sealy Gosset discovered the t-distribution and its use. Gosset, whose statistical analysis stemmed from his work as a brewer at Guiness Son & Co. in Dublin, Ireland, approached Karl Pearson for advice in 1905 and spent 1906–1907 in Pearson's Biometric Laboratory in London. The fruits of this period were two papers published in *Biometika* in 1908. The first paper, which is on what is now called Student's t-distribution, is remarkable in two respects: (1) it derived the sampling distribution of the sample variance (s^2), making the key distinction between sample and population variances that the previous literature had tended to obscure, and shifting attention from large-sample to small-sample statistics, and (2) the paper used random sampling to obtain the empirical distribution of the t-statistic in order to compare it to the theoretical result. ["The probable error of a mean," W. S. Gosset "Student," *Biometrika*, 6, 1908, 1–24; "A history of distribution sampling prior to the era of the computer and its relevance to simulation," D. Teichroew, *Journal of the American Statistical Association*, 60, 1965, 27–49; *Counting for Something: Statistical Principles and Personalities*, W. S. Peters, Springer-Verlag, New York, 1987, 100–126; *Statisticians of the Centuries*, G. C. Heyde, E. Seneta, editors, Springer-Verlag, New York, 2001]

Digital statistical testing:

Gosset's t-distribution paper used a contemporary study's data set that contained height and left-middle finger measurements of 3000 criminals. After transferring the measurements for each criminal onto a piece of cardboard, the pieces were shuffled and used to produce 750 samples of size 4.

William Sealy Gosset
1876–1937

1909 Erlang and telephone traffic

Agner Krarup Erlang was introduced to telephone system problems by J. Jensen (of Jensen's inequality), chief engineer at the Copenhagen Telephone Company. Erlang's first major publication on modeling telephone traffic showed that incoming calls can be characterized by the Poisson distribution. In a 1917 paper, he calculated the famous Erlang loss formulas. Through these writings, Erlang laid the foundations of modern queueing theory, described how to write the balance equations for states, and invented the method of phases. His concept of "statistical equilibrium," which he used to justify some ergodic results, would not pass the test of today's rigor, but allowed him to use his insights to study system behavior. ["The theory of probabilities and telephone conversations," A. K. Erlang, *Nyd tidsskrift for Matematik*, B, 20, 1909, 33; "Operational Research in some other countries," E. Jensen, *Proceedings of the First International Conference on Operational Research*, Operations Research Society of America, Baltimore, 1957; "Sixty years of queueing theory," U. N. Bhat, *Management Science*, 15, 1969, B-280–294]

Where it all began:

In his short statement on the status of operations research in Denmark 1957, Erik Jensen (1957) wrote: "I am quite the wrong person to ask to say anything about operational research in Denmark, because I do not know much about operational research. I can say, however, that for many years in Denmark it has been a one-man show. I do not know exactly when it began, but it has been carried out in the Copenhagen Telephone Company for a great many years."

Agner Krarup Erlang
1878–1929

1909 Facility location

The 17th century French mathematician, Pierre de Fermat, in his treatise on maxima and mimima, proposed in 1643 a problem that can be interpreted as a facility location problem: "Let he who does not approve of my method attempt the solution of the following problem – Given three points in a plane, find a fourth point such that the sum of its distances to the three given points is a minimum." Here it is assumed that all angles of the triangle are less that 120°. Circa 1645, the Italian mathematician and physicist Evangelista Torricelli solved the problem by showing that the circles circumscribing the equilateral triangles constructed on the sides of and outside the triangle formed by the three points intersect in the point sought (the Torricelli point). In 1750, the English mathematician, Thomas Simpson (of Simpson's Rule for numerical integration) generalized the three-point problem to find the point that minimizes a weighted sum of the distances. The German mathematician, Franz Heinen, showed (1834) that if an angle of the triangle is 120° or more, then the Torricelli point is the vertex of that angle (this case was first considered by the Italian mathematician Bonaventura Cavalieri in 1647). It was the book by the German economist Alfred Weber (1909) that brought this problem to the attention of economists and analysts as an important industrial location problem. Although he did not offer a method for solving it, Weber discusses in some detail the general problem of locating a central facility (factory, warehouse) that must send (or receive) items from several points (distribution centers, stores) such that the weighted sum of the distances to all the points is minimized. The weights are the quantities shipped between the central facility and each point. In a mathematical appendix to Weber's book, Georg Pick states: "I wish that further formulas, particularly locational figures with more than three points, might be developed." Pick discusses and pictures a mechanical device (Varignon frame) for solving the weighted three-point problem. In the 1930s, a sixteen year-old Hungarian mathematician analyzed the general problem of minimizing the distances from the facility to any number of points and proposed a method of solution. His paper, written in French and published in a Japanese mathematics journal, developed an iterative algorithm that converges to the optimal facility location under mild assumptions. This procedure was rediscovered by others in the late 1950s and 1960s. The author of the paper was Endre Weiszfeld, now known as Andrew Vazsonyi. In their book, Richard Courant and Herbert Robbins (1941) call Fermat's problem the Steiner Problem after the German mathematician Jakob Steiner. But, as Harold Kuhn (1974) points out, Steiner did not contribute anything new to the problem. [*Uber den Standort der Industrien*, A. Weber, Tübingen, 1909 (translation *Theory of the Location of Industries*, University of Chicago Press, Chicago, 1929); "Sur le point pour lequel la somme des distances de n points donnés est minimum," E. Weiszfeld, *Tôhoku Mathematical Journal*, 43, 1937, 355–386; "On a pair of dual nonlinear programs," H. W. Kuhn, pp. 39–54 in *Nonlinear Programming*, J. Abadie, editor, North-Holland, Amsterdam, 1967; " 'Steiner's' problem revisited," H. W. Kuhn, pp. 52–70 in *Studies in Optimization*, Vol. 10, G. B. Dantzig, B. C. Eaves, editors, The Mathematical Association of America, Providence, 1974; *Facilities Location*, R. Love, J. G. Morris, G. O. Wesolowsky, North-Holland, New York, 1988; "Location analysis," C. ReVelle, pp. 459–465 in *Encyclopedia of Operations Research and Management Science*, 2nd edition, S. I. Gass, C. M. Harris, editors, Kluwer Academic Publishers, Boston, 2001; *Which Door has the Cadillac: Adventures of a Real-Life Mathematician*, A. Vazsonyi, Writers Club Press, New York, 2002; "Geometrical solution to the Fermat problem with arbitrary weights," G. Jalal, J. Krarup, pp. 67–104 in *Annals of Operations Research 123, Contributions in Location Analysis: A Volume in Honor of Kenneth E. Rosing*, J. Hodgson, C. ReVelle, editors, 2003]

Evangelista Torricelli 1608–1647

Torricelli Point

Andrew Vazsonyi 1916–2003

1910 Brouwer's fixed point theorem

The Brouwer fixed point theorem states that if S is a nonempty compact convex subset of a normed linear space, then any continuous function $f : S \to S$ has a fixed point, i.e., a point x that satisfies $f(x) = x$. This topological result, due to the Dutch mathematician Luitzen E. J. Brouwer, has proved to be one of the most useful theorems in mathematical economics and game theory. In 1941, Kakutani extended it to point-to-set maps. John von

Neumann and John F. Nash both used fixed point theorems to simplify their originally more complicated proofs. Von Neumann (1937) used Brouwer's theorem to provide a topological proof of the existence of a general competitive equilibrium and, following a suggestion by David Gale, Nash (1950) based his existence proof for Nash equilibria on Kakutani's fixed point theorem. ["Über eineindeutige, stetige, Transformationen von Flächen in Sich," L. E. J. Brouwer, *Mathematische Annalen*, 67, 1910, 176–180; "Über ein Oekonomisches Gleichungssystem und eine Verallgemeinerung des Brouwerschen Fixpunktsatzes," J. von Neumann, *Ergebnisse eines Mathematischen Seminars*, K. Menger, editor, Vienna, 1937, translated as "A model of general economic equilibrium," *Review of Economic Studies*, 13, 1, 1945, 1–9; "A generalization of Brouwer's fixed point theorem," S. Kakutani, *Duke Mathematical Journal*, 8, 1941, 457–458; "Equilibrium points in n-person games," John F. Nash, Jr., *Proceedings of the National Academy of Sciences*, 36, 1950, 48–49; *Fixed Point Theorems with Applications in Economics and Game Theory*, Kim C. Borders, Cambridge University Press, Cambridge, 1985]

Luitzen E. J. (L. E. J.) Brouwer
1881–1966

Shizuo Kakutani

1912 Games with perfect information

A game with perfect information is characterized as follows: the players are aware of all the rules, possible choices, and a past history of play by all the players. Chess, backgammon, tic-tac-toe are examples of such games. The German logician, Ernst Zermelo, proved, in his paper that contained the first general theorem of game theory, that games with perfect information are strictly determined, that is, a solution exists in which both players have pure optimal strategies. ["Über eine Anwendung der Megenlehre auf die Theorie des Schachspiels," E. Zermelo, pp. 501–594 in *Proceedings of the Fifth International Congress of Mathematicians*, Vol. 2, 1912, Cambridge University Press, Cambridge, 1913; *Game Theory*, M. D. Davis, Basic Books, New York, 1970;

Ernst Zermelo
1871–1953

Games, Theory and Applications, L. C. Thomas, John Wiley & Sons, New York, 1984; "Game theory," W. F. Lucas, pp. 317–321 in *Encyclopedia of Operations Research and Management Science*, 2nd edition, S. I. Gass, C. M. Harris, editors, Kluwer Academic Publishers, Boston, 2001]

1913 Inventory Economic Order Quantity (EOQ)

The well-known square-root formula for the optimal economic order quantity (EOQ), one form of which is EOQ $= \sqrt{2KD/h}$, due to Ford W. Harris, is a cornerstone of inventory management. Here K = setup cost of placing an order, D = rate of demand for product, h = holding cost per unit. Harris published the EOQ formula in 1915. His work, however, can be traced to an earlier 1913 paper. ["How many parts to make at once," F. W. Harris, *Factory, The Magazine of Management*, 10, 2, February 1913, 135–136, 152; *Principles of Operations Research*, H. M. Wagner, Prentice-Hall, Englewood Cliffs, 1969; "An early classic misplaced: Ford W. Harris's economic order quantity model of 1915," Donald Erlenkotter, *Management Science*, 35, 7, 1989, 898–900]

1914 Lanchester's Equations

The British aeronautical engineer, Frederick W. Lanchester, who, among other things, built the first automobile in Britain, moved the study of military operations from the kriegspiel tabletop into the realm of mathematical analysis. Lanchester formulated sets of differential equations that dealt with the relationship between the concentration of forces and the effective strength of the opposing forces, the solution of which determined the expected results of a combat engagement. His analysis produced the N^2-law: The fighting strength of a force is proportional to the square of its numerical strength multiplied by the fighting value of individual units. Generalizations of Lanchester's equations have been shown to have some validity when tested against historical battles. [*Aircraft in Warfare; the Dawn of the Fourth Force*, F. W. Lancaster, Constable and Company, Ltd., London, 1916; *Methods of Operations Research*, P. M. Morse, G. E. Kimball, 1st edition revised, John Wiley & Sons, New York, 1951 (Dover reprint 2003); "A verification of Lanchester's laws," J. H. Engel, *Operations Research*, 2, 2, 1954, 163–171; "Of horseless carriages, flying machines and operations research: A tribute to Frederick William Lanchester (1868–1946)," J. F. McCloskey, *Operations Research*, 4, 2, 1956, 141–147; "Lanchester-type models of warfare," H. K. Weiss, pp. 82–99 in *Proceedings of the First International Conference on Operational Research*, M. Davies, R. T. Eddison, T. Page, editors, John Wright and Sons, Ltd., Bristol, 1957; *The War Game*, G. B. Brewer, M. Shubik, Harvard University Press, Cambridge, 1979]

Frederick W. Lanchester
1868–1946

1915 Positive solution to linear equations

Conditions for the existence of a positive solution to a set of linear equations or inequalities were investigated by the German mathematician E. Stiemke, as given in the following transposition theorem: For a matrix A, either $Ax \geqslant 0$ has a solution x or $A^T y = 0$, $y > 0$, has a solution y, but never both. ["Über positive Lösungen homogener linearer Gleichungen," E. Stiemke, *Numer. Ann.*, 76, 1915, 340–342; *Nonlinear Programming*, O. L. Mangasarian, McGraw-Hill, New York, 1969]

1920 Maximum likelihood method

The notion of estimating a parameter by maximizing an appropriate function constructed from the observations can be traced back to Daniel Bernoulli, Leonhard Euler, Johann Heinrich Lambert and Joseph-Louis de Lagrange. However, it was English statistician Ronald Aylmer Fisher who, in one of his most influential papers (1922), established a major strand of statistical reasoning by proposing the method of maximum likelihood as a general procedure for point estimation. His original conception of the method was published in a 1912 paper during his third year as an undergraduate. Fisher introduced the term "likelihood" in a 1920 paper (published in 1921) when he realized the importance of distinguishing between probability and likelihood. ["On an absolute criterion for fitting frequency curves," R. A. Fisher, *Messenger of Mathematics*, 41, 1912, 155–160; "On the 'probable error' of a coefficient of correlation deduced from a small sample," R. A. Fisher, *Metron*, 1, 4, 1921, 3–32; "On the mathematical foundations of theoretical statistics," R. A. Fisher, *Philosophical Transactions of the Royal Society of London*, A, 222, 1922, 309–368; "The history of likelihood," A. W. F. Edwards, *International Statistical Review*, 42, 1, 1974, 9–15; "Daniel Bernouilli, Leonhard Euler, and maximum likelihood," pp. 302–319 in *Statistics on The Table: The History of Statistical Concepts and Methods*, S. M. Stigler, Harvard University Press, Cambridge, Mass., 1999]

1921 Minimax strategies for two-person symmetric games

For the two-person, symmetric, zero-sum game, Émile Borel defined the game-theoretic framework and the concept of a "method of play" (strategy) as a code that determines "for every possible circumstance . . . what the person should do." He then pursued an optimal strategy and derived the minimax solutions for games with three or five possible strategies. In his later work, Borel doubted whether minimax solutions always existed. In 1928, John von Neumann proved the existence of an optimal minimax strategy for two-person zero-sum games. Later, Borel's collaborator, Jean Ville, supplied an elementary proof of this result. [Three of Borel's game theory papers have been published (translated by L. J. Savage) in *Econometrica*, Vol. 21, 1953; "Émile Borel, initiator of the theory of psychological games and its applications," M. Fréchet, *Econometrica*, 21, 1953, 95–96; "Creating a context for game theory," R. J. Leonard, pp. 29–76 in *Toward a History of Game Theory*, E. R. Weintraub, editor, Duke University Press, Durham, 1992]

1922 Sufficient condition for the Central Limit Theorem

While the statement of the Central Limit Theorem (CLT) dates back to Pierre-Simon Laplace in 1810, the first rigorous proof of it was given in 1901 by the Russian mathematician Alexander M. Liapanov, a student of Pafnuty L. Chebyshev. A drawback to this result was the requirement of finite third moments. In 1920, without being aware of Liapanov's proof, the Finnish mathematician Jarl Waldemar Lindeberg began to investigate conditions that would ensure CLT to hold. Lindeberg's published his proof of CLT using these conditions in 1922. His work ultimately led to the sufficient condition that the normed tail sum of the variances tend to zero as n goes to infinity. This *Lindeberg condition* was shown to be also necessary by William Feller in 1935. Working independently along a different direction, Paul Lévy was also led to similar conditions for the CLT, now referred to as the *Lindeberg–Lévy conditions.* In operations research, these conditions are used when proving asymptotic results in the analysis of algorithms or simulation techniques. ["Eine neue Herleitung des Exponentialgesetzes in der Wahrscheinlichkeitsrechnung," J. W. Lindeberg, *Math. Zeitsch.*, 1922; *The Life and Times of the Central Limit Theorem*, W. J. Adams, Kaedmon, New York, 1974; *The History of Mathematics in Finland 1828–1918*, G. Elfving, *Societas Scientarium Fennica*, Helsinki, 1981; "The Central Limit Theorem around 1935," L. Le Cam, *Statistical Science*, 1, 1986, 78; *The Lady Tasting Tea*, D. Salsburg, W. H. Freeman & Co., New York, 2001]

1925 Random digits generation

Random digits were first systematically generated by Leonard H. C. Tippett to confirm the results of his 1925 paper on extreme value distributions. Tippett sampled 5000 randomly drawn observations with replacement from a bag containing 1000 cards. The numbers on the cards followed a known normal distribution. He also used 40,000 digits from the areas of parishes recorded in the British census returns and combined them by fours to get 10,000 numbers drawn from 0000 to 9999 at random. Tippett published his list of random digits in 1927. Referring to this book, Edward U. Condon is said to have introduced Tippett by remarking that he had written a book that really could have been written by a monkey! Curiously, Tippett's random digits were initially used without any statistical tests of their randomness. The Indian statistician Prasanta Chandra Mahalanobis applied a series of randomness tests to this data and concluded that Tippett's digits were random. Since 1927, a number of other tables of random digits have been published, including RAND's well-known table of a million random digits described by Brown (1951). In the 1940s, John von Neumann, as part of his and Stanislaw Ulam's application of Monte Carlo methods to atomic bomb research at Los Alamos, developed one of the first arithmetic methods for producing pseudorandom numbers, the middle-square method, as well as statistical tests for checking sequences for random properties. ["On the extreme individuals and the range of samples taken from a normal population," L. H. C. Tippett, *Biometrika*, 17, 1925, 364–87; "Random sampling numbers," L. H. C. Tippett, *Tracts for Computers*, Vol. XV, Cambridge University Press, Cambridge, 1927; "Tables of random samples from a normal distribution," P. C. Mahalanobis, *Sankhya*, 1, 1934, 289–328; "History of RAND's random digits: Summary," W. G. Brown, pp. 31–32 in *Monte Carlo Method*, A. S. Householder, G. E. Forsythe, H. H. Germond, editors, Applied Mathematics Series, Vol. 12, U.S. National Bureau of Standards, Washington, DC, 1951; "Random number generators," T. E. Hull, A. R. Dobell, *SIAM Review*, 4, 3, 1962, 230–254; "The test-passing method of random digit selection," F. Gruenberger, *Software Age*, June 1970, 15–16; *A Million Random Digits with 100,000 Normal Deviates*, The RAND Corporation, The Free Press, New York, 1955; *Counting for Something: Statistical Principles and Personalities*, W. S. Peters, Springer-Verlag, New York, 1987, 140–141; "The transformation of numerical analysis by the computer: An example from the work of John von Neumann," W. Aspray, pp. 307–322 in *The History of Modern Mathematics, Vol. II: Institutions and Applications*, D. E. Rowe, J. McCleary, editors, Academic Press, Boston, 1989; *Monte Carlo: Concepts, Algorithms and Applications*, G. S. Fishman, Springer-Verlag, New York, 1995]

1925 ***Statistical Methods for Research Workers***, Ronald A. Fisher, Oliver and Boyd, London

This book by the celebrated English statistician and geneticist Ronald Aylmer Fisher covers his statistical activities at the Rothamsted Experimental Station for agricultural research. Fisher's objective in writing it "... is to put into the hands of research workers, and especially of biologists, the means of applying statistical tests accurately to numerical data accumulated in their own laboratories or available in the literature." Chapters discuss diagrams, distributions, tests of goodness of fit, tests of significance, the correlation coefficient, interclass correlations and analysis of variance. The book has had 13 English editions. It is of interest to quote the following from Fishers's preface: "Daily contact with the statistical problems which present themselves to the laboratory worker has stimulated the purely mathematical researchers upon which are based the methods here presented. Little experience is sufficient to show that the traditional machinery of statistical processes is wholly unsuited to the needs of practical research. Not only does it take a cannon to shoot a sparrow, but it misses the sparrow. The elaborate mechanism built on the theory of infinitely large samples is not accurate enough for simple laboratory data. Only by systematically tackling small sample problems on their merits does it seem possible to apply accurate tests to practical data. Such at least is the aim of this book." ["Fisher, R. A.," M. S. Bartlett, pp. 352–358 in *International Encyclopedia of Statistics*, W. H. Kruskal, J. M. Tanur, editors, The Free Press, New York, 1978]

Quotable Fisher:

"The science of statistics is essentially a branch of Applied Mathematics and may be regarded as mathematics applied to observational data."

"Statistical methods are essential to social studies, and it is principally by the aid of such methods that these studies may be raised to the rank of science. This particular dependence of social studies upon statistical methods *has led to the painful misapprehension that statistics is to be regarded as a branch of economics, whereas in truth economists have much to learn from their scientific contemporaries, not only in general scientific method, but in particular in statistical practice* (emphasis added)."

Ronald A. Fisher
1890–1962

1926 Subjective probability

The notion of degrees of belief, which is linked with the topic now called subjective probability, dates back to the earliest investigations of Jakob Bernoulli I (*Ars Conjectandi*, 1713), and was pursued by Émile Borel, John Venn, and John Maynard Keynes, among others. Frank P. Ramsey believed that the only way to measure degrees of belief is to observe overt behavior manifested in choices. He thus linked subjective probability with the concept of utility and explicit choices. His famous "Truth and Probability" paper was written in 1926. Ramsey also introduced the notion of *coherence* to require conformance to the laws of probability. With Bruno de Finetti's concept of *exchangeable events* and Ramsey's derivation of the limiting distribution of relative frequency for such events (1930), the connection between subjective and classical probability was made in a rigorous fashion. ["Truth and Probability," F. P. Ramsey, Chapter 7 in *The Foundation of Mathematics and other Logical Essays*, R. B. Braithwaite, editor, The Humanities Press, New York, 1950; "Funzione caratteristica di un fenomeno aleatorio," B. de Finetti, *Memorie della Academia dei Lincei*, 4, 1930, 86–133; "Foresight: Its logical laws, its subjective sources," B. de Finetti, pp. 93–158 in *Studies in Subjective Probability*, H. E. Kyburg, Jr., H. E. Smokler, editors, John Wiley & Sons, New York, 1964; *Creating Modern Probability*, J. von Plato, Cambridge University Press, Cambridge, 1994]

1927 Applications of probability theory to telephone engineering

Edward C. Molina, a self-taught researcher, made seminal contributions to telephone traffic theory. The first automatic telephone exchange had been installed at La Porte, Indiana in 1892 and gave rise to the problem of expanding exchanges. There were early attempts to use probability theory for the analysis of exchanges. Starting in 1908, Molina extended this work and obtained new results that were widely used. Molina analyzed the $M/M/n$ queueing model by means of birth-and-death processes. Thornton C. Fry, a valuable contributor to the subject, organized the literature into a comprehensive theory. His 1928 book on congestion theory became the classic text on the subject. ["Application of the theory of probability to telephone trunking problems," E. C. Molina, *Bell Systems Technical Journal*, 6, 1927, 461–494; *Probability and its Engineering Uses*, T. C. Fry, Van Nostrand, New York, 1928; *Introduction to Congestion Theory in Telephone Systems*, R. Syski, Oliver and Boyd, Edinburgh, 1960; "Sixty years of queueing theory," U. N. Bhat, *Management Science*, 15, 1969, B-280–294]

1927 Statistical analysis of time series

The use of combined autoregressive moving average processes (ARMA) for studying time series was suggested by the British statistician George Udny Yule and the Russian economist and statistician Eugene Slutsky. They observed that starting with a series of purely random numbers, one can take sums or differences of such numbers to produce new series that exhibit the cyclic properties often seen in time series. This work laid the foundation of autoregressive integrated moving average (ARIMA) models proposed by George E. P. Box and Gwilym M. Jenkins nearly 45 years later. Slutsky and Yule are also remembered for the "Slutsky–Yule Effect" which states that a moving average of a random series may exhibit oscillatory movement when none existed in the original data.

["On a method for investigating periodicities in disturbed series with special reference to Wölfer's sunspot numbers," G. U. Yule, *Philosophical Transactions of the Royal Society London*, A, 226, 1927, 267–298; "Autoregressive and moving-average time-series processes," M. Nerlove, F. X. Diebold, pp. 25–35 in *The New Palgrave: Time Series and Statistics*, J. Eatwell, M. Milgate, P. Newman, editors, W. W. Norton & Co., New York, 1990; *Statisticians of the Centuries*, G. C. Heyde, E. Seneta, editors, Springer-Verlag, New York, 2001]

1928 Existence proof for an equilibrium strategy for two-person matrix games

Two publications by John von Neumann appeared in 1928 dealing with the minimax proof for two-person matrix (zero-sum) games. The first was a communication to É. Borel, in which von Neumann announced that he had solved the problem of finding an optimal strategy for the two-person, zero-sum game. The second contained a long and difficult existence proof for the equilibrium of the two-person, discrete game. It also included two examples of zero-sum games with only mixed strategy solutions. Von Neumann had already produced a proof in 1926 and presented it in shorter form to the Göttingen Mathematical Society in December 1926. Twenty-five years later, George B. Dantzig showed how linear programming provides a constructive proof for finding the solution to any two-person matrix game. ["Zur Theorie der Gesellschaftsspiele," J. von Neumann, *Mathematische Annalen*, 100, 1928, 295–320, translated as "On the theory of games of strategy," pp. 13–42 in *Contributions to the Theory of Games*, A. W. Tucker, R. D. Luce, editors, Princeton University Press, Princeton, 1959; "A proof of the equivalence of the programming problem and the game problem," G. B. Dantzig, pp. 330–355 in *Activity Analysis of Production and Allocation*, T. C. Koopmans, editor, John Wiley & Sons, New York, 1951; "Émile Borel, initiator of the theory of psychological games and its applications," M. Fréchet, *Econometrica*, 21, 1953, 95–96; "Creating a context for game theory," Robert J. Leonard, pp. 29–76 in *Toward a History of Game Theory*, E. R. Weintraub, editor, Duke University Press, Durham, 1992]

Two persons do not equal zero-sum:

Maurice Fréchet (1953) argued that Borel should get credit as the originator of modern game theory; von Neumann's response was that Borel did not prove the general theorem.

Émile Borel
1871–1956

John von Neumann
1903–1957

1929 Sequential sampling procedure

For a given lot of items (e.g., manufactured parts), acceptance sampling involves drawing a random sample and accepting the lot if the sample contains less than a specified number of defective units. The sampling can be exhaustive, whereby all items are examined, but this is usually very costly and time consuming. Typically, the sample size is a fraction of the lot size. The idea of sequential sampling, which calls for the drawing of a second sample based on the analysis of a first sample, was due to Harold F. Dodge and Harold G. Romig of Western Electric. The advantage of this two-stage process is that, on the average, it reduces the total sample size as compared to one-stage sampling. ["A method of sampling inspection," H. F. Dodge, H. G. Romig, *The Bell System Technical Journal*, 8, 1929, 613–631; *Quality Control and Industrial Statistics*, 4th edition, A. J. Duncan, Richard D. Irwin, Homewood, 1974]

1929 Characterization of planar graphs

Kazimierz Kuratowski
1896–1980

The Polish mathematician Kazimierz Kuratowski showed that if a graph is non-planar it must contain either the complete graph on 5 nodes (K_5) or the bipartite graph on 6 nodes ($K_{3,3}$) as subgraphs. This result was announced to the Polish mathematical society in Warsaw on June 21, 1929. The graph $K_{3,3}$ is the subject of the well-known "water, gas, electricity" (three houses connected to three utilities) problem that was known to be non-planar much earlier. Planar graphs are often investigated when more general network algorithms are specially designed to handle planar graphs. ["Sur le problème des courbes gauches en topolgie," K. Kuratowski, *Fundamenta Mathematicae*, 15, 1930, 271–283, extract reprinted in *Graph Theory 1736–1936*, N. L. Biggs, E. K. Lloyd, R. J. Wilson, Oxford University Press, Oxford, 1976; *Graphs as Mathematical Models*, G. Chartrand, Prindle, Weber & Schmidt, Boston, 1977]

1930 Confidence limits

The implicit use of confidence limits to provide a range of possible values for estimated parameters can be traced back to Laplace and Gauss, but, since the limits derived were approximations, the underlying logic of the procedure remained obscure. Ronald A. Fisher (1930) was the first to recognize that such limits can be justified and given an exact meaning without appeal to *a priori* probabilities or Bayesian priors. Fisher used the expression "fiducial probability" to refer to confidence statements. The importance of this advance was duly recognized and highlighted by Jerzy Neyman (1934) (originally written in Polish in 1933). Neyman introduced the terminology of confidence limits. ["Inverse probability," R. A. Fisher, *Proceedings of the Cambridge Philosophical Society*, 26, 1930, 528–535, also, pp. 194–201, David and Edwards (2001); "On the two different aspects of the representative method," J. Neyman, *Journal of the Royal Statistical Society*, 97, 1934, 558–625; *Annotated Readings in the History of Statistics*, H. A. David, A. W. F. Edwards, Springer-Verlag, New York, 2001]

1930 The Econometric Society founded

The Econometric Society, an international society for the advancement of economic theory in its relation to statistics and mathematics, was founded in 1930. Many of its members have made seminal contributions to operations research, and some important theoretical and applied OR papers have appeared in its flagship journal *Econometrica*. Its first president was Irving Fisher. [http://www.econometricsociety.org/thesociety.html]

1930 Pollaczek formula for $M/G/1$ queues

Félix Pollaczek was a pioneer in the study of queueing systems. He developed the formula for a customer's mean waiting time in an $M/G/1$ queueing system. The formula was derived independently a few years later by Alexander Khintchine, and it is now known as the Pollaczek–Khintchine formula. If W_q is the mean waiting time in a queue with Poisson arrivals at the rate of λ and a general service time with mean $E(S)$ and variance Var(S), the formula states $W_q = [\rho^2 + \lambda^2\text{Var}(S)]/[2\lambda(1-\rho)]$, where $\rho = \lambda E(S)$. In his subsequent work, Pollaczek studied the $GI/G/1$ and $GI/G/s$ systems extensively and came to view the latter as a very hard problem. ["Über eine Aufgabe der Wahrscheinlichkeitstheorie," F. Pollaczek, *Mathematische Zeitschrift*, 32, 1930, 64–100; *Statisticians of the Centuries*, G. C. Heyde, E. Seneta, editors, Springer-Verlag, New York, 2001]

1931 Quality control charts

Walter A. Shewhart joined Western Electric Company in 1918 and was transferred to the Bell Telephone Laboratories in 1925. He remained there until his retirement in 1956. In the early 1920s, work on control charts started at Western Electric as part of a company-wide view of quality assurance based on scientific principles. In addition to Shewhart, the quality assurance team included Harold F. Dodge, Thornton C. Fry, Edward C. Molina, and Harold G. Romig. In 1931, Shewhart published his major work on control charts that set the direction for the entire field. His work developed the key concepts of assignable causes and that of a system being under statistical control. Shewhart was heavily committed to the broader scientific methodology underlying statistical quality control which involved a view of the organization and its uses of measurement (the Shewhart cycle). W. Edwards Deming closely followed Shewhart's work and was instrumental in getting Shewhart's 1939 book published. [*Economic Control of Quality of Manufactured Products*, W. A. Shewhart, D. Van Nostrand Company, New York, 1931 (republished by the American Society for Quality Control, 1980); *Statistical Method from the Viewpoint of Quality Control*, W. A. Shewhart, The Graduate School, U.S. Department of Agriculture, Washington, 1939 (Dover reprint 1986); *Statisticians of the Centuries*, G. C. Heyde, E. Seneta, editors, Springer-Verlag, New York, 2001; "Quality Control," F. Alt, K. Jain, pp. 661–674 in *Encyclopedia of Operations Research and Management Science*, 2nd edition, S. I. Gass, C. M. Harris, editors, Kluwer Academic Publishers, Boston, 2001]

Walter A. Shewhart
1891–1967

1931 The König–Egerváry theorem

In every bipartite graph, the maximum cardinality matching and the minimum node cover have the same size. This classical result is known as the König–Egerváry theorem and is one of the first examples of a combinatorial min-max relationship. Subsequent relations of this form include the minimum-cut maximum-flow theorem and Edmonds's results on matching. ["Graphen und Matrizen," D. König, *Matematika és Fizikai Lápok*, 38, 1931, 116–119; "Matrixok kombinatororikus tulajdonságairól," J. Egerváry, *Matematika és Fizikai Lápok*, 38, 1931, 16–28; "Polyhedral Combinatorics," W. R. Pulleyblank, pp. 371–446 in *Handbooks in Operations Research & Management Science, Vol. 1: Optimization*, G. L. Nemhauser, H. G. Rinnooy Kan, M. J. Todd, editors, North-Holland, New York, 1989]

1931 Chapman–Kolmogorov equations

Prior to 1930, random processes studied in probability theory generally used a discrete time parameter. This changed with the publication of the "Analytical Methods" paper of Andrei N. Kolmogorov (1931) on continuous-time random processes. Together, their results laid the foundation of continuous-time Markov processes. Consider a Markov process with continuous time parameter and a countable number of states. Let $P_{in}(r,t)$ be the conditional probability of finding the process in state n at time t, given that it was in state i at some previous time $r < t$. Then, for some intermediate time s between r and t, P satisfies the Chapman–Kolmogorov equation: $P_{in}(r,t) = \sum P_{ik}(r,s)P_{kn}(s,t)$, with side conditions $\sum P_{ik}(r,t) = 1$, and $P_{ik}(r,t) \geqslant 0$, where the summations run over all states k. From this fundamental relation, one can derive systems of differential equations for $P_{in}(r,t)$, known as the forward and backward Chapman–Kolmogorov differential equations. The physicist Sidney Chapman had derived a version of these equations in 1928, while studying the Brownian motion of grains in fluids. William Feller continued the work of Kolmogorov and studied the solutions to the system of equations for more general processes. ["On the Brownian displacements and thermal diffusion of grains suspended in a non-uniform fluid," S. Chapman, *Proceedings of the Royal Society*, A, 119, 1928, 34–60; "Über die analytischen Methoden in der Wahrscheinlichkeitsrechnung," A. N. Kolmogorov, *Mathematische Annalen*, 104, 1931, 415–458; *An Introduction to Probability Theory and its Applications*, 3rd edition, W. Feller, John Wiley & Sons, New York, 1968; *Creating Modern Probability*, J. von Plato, Cambridge University Press, Cambridge, 1994; "Andrei Nikolaevich Kolmogorov: A biographical sketch of his life and creative paths," A. N. Shirayev, pp. 1–87 in *Kolmogorov in Perspective*, History of Mathematics, Vol. 20, American Mathematical Society, Providence, RI, 2000]

The importance of being Andrei:

Shirayev (2000) cites Pavel S. Alexandrov and Alexander Khinchin on the impact of Kolmogorov's "Analytical Methods" paper: "In the whole of probability theory in the twentieth century it would be hard to find another investigation that has been so fundamental for the further development of the science and its applications as this paper of Andrei Nikolaevich. In our day it has led to the development of an extensive area of study in probability: the theory of random processes .... The differential "Kolmogorov equations" that govern Markov processes and that have been mathematically grounded rigorously ..., ... contained as special cases all the equations that up to that time had been derived and applied by physicists for isolated reasons, by rule-of-thumb methods ... without any clear explanation of the premises on which they were based."

Andrei N. Kolmogorov
1903–1987

1932 Hypothesis testing

During the period 1926–1933, Jerzy Neyman and Egon S. Pearson developed the theory of hypothesis testing in response to Ronald A. Fisher's ad hoc approach. Their theory allowed one to identify optimal tests by specifying the alternative hypothesis and recognizing the basic two types of error. The celebrated Neyman–Pearson lemma, which dates back to 1930, became the fundamental tool of hypothesis testing and was seminal to advances in the later development of mathematical statistics. Karl Pearson presented the Neyman–Pearson paper, "On the problem of the most efficient tests of statistical hypotheses," to the Royal Society on November 11, 1932. The "big paper" was published in the Society's *Philosophical Transactions* the following year. [*Neyman – from Life*, Constance Reid, Springer-Verlag, New York, 1982; "Egon Sharpe Pearson," F. N. David, pp. 650–652 in *Encyclopedia of Statistical Sciences*, Vol. 6, S. Kotz, N. L. Johnson, editors, John Wiley & Sons, New York, 1982]

Egon Sharpe Pearson
1895–1980

E. S. P.:

Egon Sharpe Pearson was the only son of Karl Pearson. He was a founding member of the British Operational Research Club, the percursor of the Operational Research Society.

Jerzy Neyman
1894–1981

1933 Birth of mathematical statistics

Stephen M. Stigler, an historian of statistics, selects 1933 as a point estimate for the birth of mathematical statistics. He clarifies that this date does not refer to the birth of the various concepts that make up the subject (many of these date back to earlier centuries), but to the "birth of mathematical statistics as a *discipline*." Institutionally, Stigler notes that Harry C. Carver founded the *Annals of Mathematical Statistics* in 1930 "loosely under the aegis of the American Statistical Association (ASA)." However, in 1933, ASA cut its affiliation with this journal. Carver and a group of mathematical statisticians then formed the Institute of Mathematical Statistics (IMS) on September 12, 1935, with H. L. Rietz as president and Walter Shewhart as vice-president. The *Annals* was designated as the official journal of IMS. ["The history of statistics in 1933," pp. 157–172 in *Statistics on the Table: The History of Statistical Concepts and Methods*, S. M. Stigler, Harvard University Press, Cambridge, 1999]

1933 Principal components analysis

Although the method of principal components dates back to Karl Pearson (1901), the general procedure is due to the pioneering paper of Harold O. Hotelling (1933), a professor of economics at Columbia University. Principal components are a sequence of uncorrelated linear combinations of the original measurements, each with a variance smaller than the previous one, that collectively preserve the total variation of the original measurements. Hotelling showed how these components can be found from the eigenvectors of the population covariance matrix. ["On lines and planes of closest fit to systems of points in space," Karl Pearson, *Philosophical Magazine*, B, 2, 1901, 559–572; "Analysis of a complex of statistical variables into principal components," H. O. Hotelling, *Journal of Educational Psychology*, 24, 1933, 417–441, 498–520; *A User's Guide to Principal Components*, J. Edward Jackson, John Wiley & Sons, New York, 1991]

Harold O. Hotelling
1895–1973

1933 *Grundbegriffe der Wahrscheinlichkeitsrechnung*, Andrei Kolmogorov, Fasc. 3 of Vol. 2 of *Ergebnisse der Mathematik*, Berlin; English version, *Foundations of the Theory of Probability*, Chelsea, New York, 1950

In this celebrated book, Andrei Kolmogorov provided the axiomatic development of probability theory in terms of measure theory. This book became the symbol of modern probability theory, superseding all earlier approaches. An important new development was the treatment of stochastic processes. [*Creating Modern Probability*, J. von Plato, Cambridge University Press, Cambridge, 1994]

1935 Martingales

Paul Lévy's investigations of the abstract unifying concepts of probability theory led him to a sequence of random variables where expectation of the next variable in the sequence is always equal to the value of the last one. Lévy used the term *martingale* for such a sequence. This term referred to a device used by French farmers to keep a horse's head down and to keep the animal from rearing. By 1940, martingales became important tools in mathematical probability theory, with further theoretical results developed by J. L. Doob. [*Stochastic Processes*, J. L. Doob, John Wiley & Sons, New York, 1953; "Harnesses," J. M. Hammersley, *Proceedings of the Fifth Berkeley Symposium on Mathematical Statistics and Probability*, 3, 1966, 89–117; *The Lady Tasting Tea*, D. Salsburg, W. H. Freeman & Co., New York, 2001]

1935 Matroids

In his classic paper, the mathematician Hassler Whitney introduced the axioms for an algebraic structure he called matroids. A matroid M is a finite set S and a collection $\boldsymbol{F}$ of subsets of S, called independent sets, which play a role analogous to bases for a vector space. The axioms require that for any member X of $\boldsymbol{F}$, all proper subsets of X (including $\emptyset$) are also members of $\boldsymbol{F}$, and for two members X and Y of $\boldsymbol{F}$ of cardinality r and $r+1$, respectively, there is an element of $(Y-X)$ such that its addition to X produces a member of $\boldsymbol{F}$. Interest in matroids, especially their connections to graph theory, networks, combinatorial optimization, and greedy algorithms, was revived in the 1950s and 1960s by W. T. Tutte, Jack Edmonds, and others. Of significant influence were the papers given at a special "Seminar in Matroids," held at the National Bureau of Standards, Washington, DC, August 31–September 11, 1964. ["On the abstract properties of linear dependence," H. Whitney, *American Journal of Mathematics*, 57, 1935, 509–533; "Lectures on Matroids," W. T. Tutte, *Journal of Research of the National Bureau of Standards*, 69B, 1965, 1–48; "Matroids and the greedy algorithm," J. Edmonds, *Mathematical Programming*, 1, 1971, 127–137; *Combinatorial Optimization: Networks and Matroids*, Eugene L. Lawler, Holt, Rinehart and Winston, 1976]

1935 *The Design of Experiments*, Ronald A. Fisher, Oliver & Boyd, Edinburgh

This classic book summarizes Fisher's path-breaking work in the design of experiments. It is well-known for its celebrated illustration of the lady tasting tea. During 1924–1926, Fisher developed such basic principles of experimental design as factorial designs,

Latin squares, confounding, and partial confounding, and the analysis of covariance. Fisher is regarded as the father of modern statistics. [*R. A. Fisher: The Life of a Scientist*, J. Fisher Box, John Wiley & Sons, New York, 1978; *Encyclopedia of Statistical Sciences*, Vol. 2, S. Kotz, N. L. Johnson, editors, John Wiley & Sons, New York, 1982; *Statisticians of the Centuries*, G. C. Heyde, E. Seneta, editors, Springer-Verlag, New York, 2001]

3

Birth of operations research from 1936 to 1946

1936 Time Zero: British military applications of OR

The birth date of operations research (or operational research, its British natal name) cannot be stated unequivocally. The year 1936 was the year the British Air Ministry established the Bawdsey Manor Research Station, Suffolk, to study how newly developed radar technology could be used for controlled interception of enemy aircraft. Bawdsey was first directed by Robert Watson-Watt, superintendent of the Radio Department of the National Physical Laboratory. The efforts of a team of RAF officers and civilian scientists, working in 1936 at Biggin Hill Airfield in Kent, is widely considered to be the embryonic and seminal applied research activity that set in motion what was soon to be called operational research. The British Operational Research Society, which celebrated 50 years of OR in 1987, set OR's origins in 1937 (about the time it was clear that the Bawdsey scientists' radar deployment studies would be of value in the defense of Britain). The term operational research is attributed to A. P. Rowe (who superseded Watson-Watt as superintendent of the Bawdsey Research Station), when, in 1938, he had teams from Bawdsey examine the efficiency of the plotting and operations room technique that originated from the Biggin Hill radar interception experiments. In 1939, these teams were made part of the Operational Research Section. In 1941, it became the Operational Research Section, RAF Fighter Command. ["Division of social and international relations of science report of the Dundee meeting. August 30, 1947," British Association, 1947, reprinted as "Operational research in war and peace," *The Advancement of Science*, 17, 1948, 320–332; "H. J. Larnder," pp. 3–12 in *Proceedings of the Eighth IFORS International Conference on Operational Research*, K. B. Haley, editor, North-Holland, 1979; "Fifty years of operational research," J. Rosenhead, *Journal of the Operational Research Society*, 38, 1, 1987, 1; "Reminiscences of operational research in World War II by some of its practitioners," F. L. Sawyer, A. Charlesby, T. E. Easterfield, E. E. Treadwell, *Journal of the Operational Research Society*, 40, 2, 1989, 115–136; "Air defence of Great Britain, 1920–1940: An operational research perspective," M. Kirby, R. Capey, *Journal of the Operational Research Society*, 48, 6, 1997, 555–568; *Operational Research in War and Peace*, M. W. Kirby, World Scientific, London, 2003]

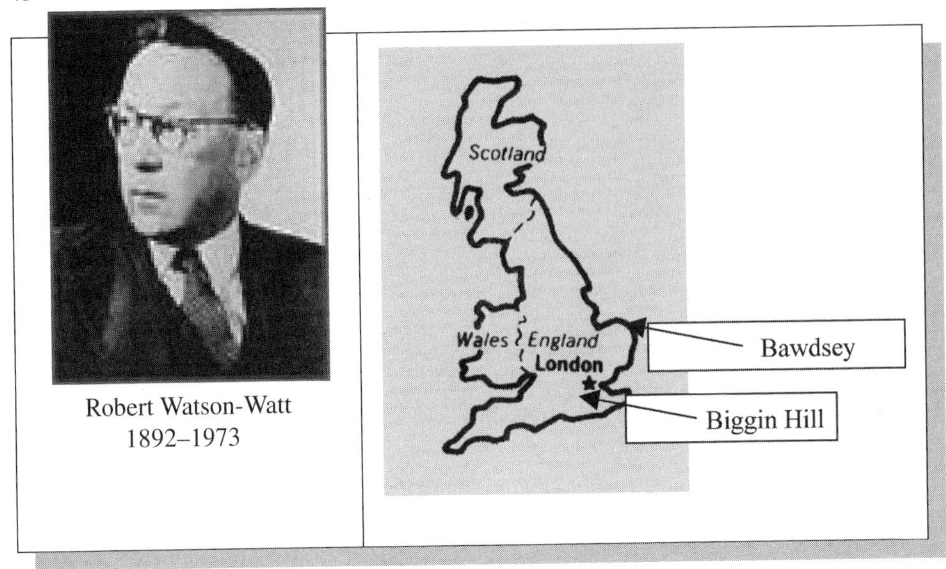

Robert Watson-Watt
1892–1973

1936 Solution of linear inequalities

Theodore S. Motzkin
1908–1970

Prior to 1936, there were few papers dealing with the solution of sets of linear inequalities. The 1936 dissertation of the German mathematician Theodore S. Motzkin cited only 42 such papers. Motzkin's Transposition Theorem for linear inequalities is a more general form from which Gordan's and Stiekme's transposition theorems can be derived. It also can be used to prove the duality theorem of linear programming. [*Beiträge zur Theorie der Linearen Ungleichungen*, T. S. Motzkin, Doctoral Thesis, University of Zurich, 1936; *Linear Programming and Extensions*, G. B. Dantzig, Princeton University Press, Princeton, 1963]

1936 Interindustry economics

With the publishing of his first table (matrix) of input–output coefficients in 1936, Wassily W. Leontief, a Russian born economist, who had recently joined the faculty of Harvard University, established the field of interindustry economics. For an economy, the coefficients show the amount (input) of one industry required to produce one unit (output) of each of the economy's industries. Although Leontief's matrix assumes linearity (input and output are proportional) and is non-dynamic, applications of interindustry (input–output) economics to analyze the impact of a government's economic policy and changes in consumer activity have proven to be of great value; it has been used by the U.S. Department of Labor Statistics, the World Bank and the United Nations.

Leontief received the 1973 Nobel prize in economics for the development of the input–output method and for its application to important economic problems. ["Quantitative input and output relations in the economic system of the United States," W. W. Leontieff, *Review of Economic Statistics*, 18, 1936, 105–125; *The Structure of American Economy, 1919–1929*, W. W. Leontief, Harvard University Press, Cambridge, 1941; *The Structure of American Economy, 1919–1939*, 2nd edition, W. W. Leontief, Oxford University Press, Oxford, 1951; *Linear Programming and Extensions*, G. B. Dantzig, Princeton, 1963; `http://www.econlib.org/Enc/bios/Leontief.html`; `http://www.garfield.library.upenn.edu/essays/v9p272y1986.pdf`]

Linear programming precursor:

George B. Dantizg (1963), the inventor of linear programming, cites Leontief's interindustry structure as a motivating factor in Dantzig's development of the general linear-programming model.

(©Nobel Foundation)
Wassily Leontief
1906–1999

1936 Turing machines

In the course of his program on the foundations of mathematics, David Hilbert asked the question: Is there a fixed procedure capable of deciding whether a mathematical assertion is true for every mathematical assertion that can be formally stated? This question, called the decision problem (*Entscheidungsproblem*), attracted the attention of Alan M. Turing in 1935 when he was an undergraduate at King's College, Cambridge. In 1936, he wrote the celebrated paper that answered the question in the negative. In this paper, Turing formalized the notion of computability and introduced the Turing machine as a model for a universal computing machine. ["On computable numbers, with an application to the *Entscheidungsproblem*," A. M. Turing, *Proceedings of the London Mathematical Society (2)*, 42, 1937, 230–265; *John von Neumann and the Origins of Modern Computing*, W. Aspray, MIT Press, Cambridge, MA, 1990; *Alan Turing: The Enigma*, A. Hodges, Walker and Company, New York, 2000]

The basic computer:

A Turing machine consists of (1) a control unit, which can assume any one of a finite number of possible states; (2) a tape, marked off into discrete squares, each of which can store a single symbol taken from a finite set of possible symbols; and (3) a read–write head, which moves along the tape and transmits information to and from the control unit. The concept of a Turing machine provides the formal basis of subsequent work in complexity theory, including the definition of the classes P and NP.

Alan M. Turing
1912–1954

1936 *Theorie der endlichen und unendlichen Graphen*, Dénes König, M. B. H., Leipzig, 1936 (Chelsea Publishing Co. reprint 1950)

Two hundred years after Euler's pioneering work on the Königsberg Bridge problem, König's work introduced the term graph theory and provided the first comprehensive treatment of the subject, establishing it as a subfield of mathematics.

1937 The traveling salesman problem

Merrill M. Flood is credited with popularizing this most celebrated combinatorial problem: A traveling salesman wants to visit each of n cities exactly once and then return to his home city; if the distance (cost) of traveling from city i to city j is c_{ij}, what route (tour, circuit) should the salesman take to minimize the total distance traveled? While Flood's paper, "The traveling-salesman problem," appeared in *Operations Research* in 1956, its history seems to go back considerably more. Flood recalls being told about the problem by Albert W. Tucker in 1937, while Tucker recalled that Hassler Whitney was his possible source around 1931–1932. Flood continued to promote the problem in the late 1940s; John Williams urged him to make it known at the RAND Corporation to create other intellectual challenges besides game theory. Earlier statements of the traveling salesman problem (TSP), as a path rather than a tour, can also be traced to Karl Menger as finding the shortest polygonal graph joining a set of points in the context of defining the curve length. Menger called this the messenger problem "... because in practice the problem has to be solved by every postman, and also by many travelers." Flood was president of TIMS in 1955. ["Das Botenproblem," K. Menger, *Kolloquium*, 9, 1932, 12; "Solution of a large-scale traveling-salesman problem," G. Dantzig, R. Fulkerson, S. Johnson, *Journal of the Operations Research Society of America*, 2, 4, 1954, 393–410; "The traveling-salesman problem," M. M. Flood, *Journal of the Operations Research Society of America*, 4, 1, 1956, 61–75; "History," A. J. Hoffman, P. Wolfe, Chapter 1 in *The Traveling Salesman Problem*, E. L. Lawler, J. K. Lenstra, A. H. G. Rinnooy Kan, D. B. Shmoys, editors, John

Wiley & Sons, New York, 1985; "Traveling salesman problem," K. L. Hoffman, M. Padberg, pp. 849–853 in *Encyclopedia of Operations Research and Management Science*, 2nd edition, S. I. Gass, C. M. Harris, editors, Kluwer Academic Publishers, Boston, 2001]

Merrill M. Flood
1908–1991

See the USA in a Chevrolet:

This an optimal traveling-salesman tour of the lower 48 United States and the District of Columbia (Dantzig et al., 1954). The tour length is 12,345 miles.

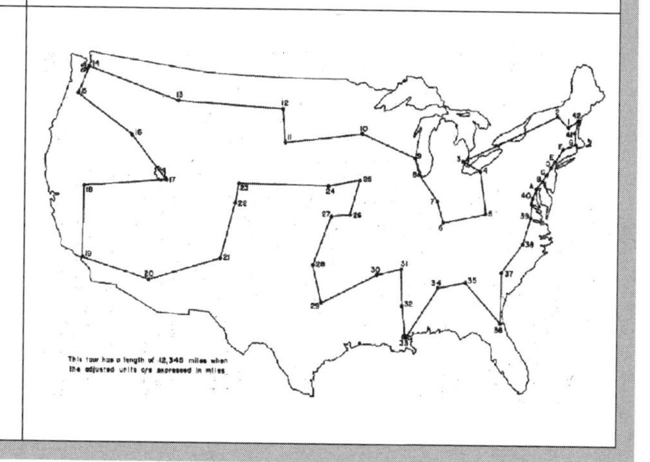

1939 Nonlinear programming optimality conditions

As part of his masters thesis in the department of mathematics, University of Chicago, William Karush stated optimality conditions for nonlinear programs, a result identical to the well-known but subsequent Kuhn–Tucker statement (1951). Karush's work was not published and remained unnoticed for many years. The optimality conditions are now referred to by the joint name of Karush–Kuhn–Tucker (KKT conditions). [*Minima of Functions of Several Variables with Inequalities as Side Conditions*, W. Karush, MSc Thesis, Department of Mathematics, University of Chicago, Chicago, 1939; "Nonlinear programming," H. W. Kuhn, A. W. Tucker, pp. 481–492 in *Proceedings of the Second Berkeley Symposium on Mathematical Statistics and Probability*, J. Neyman, editor, University of California Press, Berkeley, 1951]

1939 *Mathematical Methods of Organization and Planning Production*, L. V. Kantorovich, Leningrad State University

In this monograph, the Russian mathematician/economist, Leonid V. Kantorovich, gave a mathematical description of a production assignment problem that can be interpreted as the first statement of such a problem as a linear program. He also proposed a computational procedure (resolving multipliers) for solving it, and, further, noted that his mathematical structures could be used to analyze problems in oil refining, utilization of fuel types, minimization of scrap, construction planning, the distribution of freight over a network, and the optimum distribution of arable land to different agricultural crops. For an English translation, see *Management Science*, 4, 1960, 266–422. Based on this and subsequent publications, Kantorovich established himself as a pioneer in both the theory and application of linear programming. [*Linear Programming and Extensions*, G. B. Dantzig, Princeton, 1963; "Academician L. V. Kantorovich (19 January 1912 to 7 April 1986)," V. L. Makarov, S. L. Sobolev, pp. 1–7 in *Functional Analysis, Optimization, and Mathematical Economics: A Collection of Papers Dedicated to the Memory of Leonid Vital'evich Kantorovich*, L. J. Leifman, editor, Oxford University Press, New York, 1990]

The USSR vs. capitalistic linear programming:

As noted by Makarov and Sobolev (1990): "Unfortunately, due to the well-known peculiarities of the USSR economy in those years, as well as the absence of computers, Kantorovich's works on linear programming did not find wide enough application at that time and actually remained unknown not only abroad but also in the USSR."

(©Nobel Foundation)
Leonid V. Kantorovich
1912–1986

1940 Blackett's "Circus"

Under the direction of the physicist Patrick M. S. Blackett, a multidisciplinary group (three physiologists, one general physicist, two mathematical physicists, two mathematicians, one astrophysicist, one Army officer, and one surveyor) was assembled under the formal organizational title of The Anti-Aircraft Command Research Group, Royal Air Force, to study the use of radar in anti-aircraft gunnery. Known as the Blackett's Circus, it established the concept of an OR multidisciplinary team and demonstrated the value and effectiveness of such teams when applied to complex, real-world problems. ["A history of Operations Research," F. N. Trefethen, pp. 3–35 in *Operations Research for Management*, J. F. McCloskey, F. N. Trefethen, editors, The John Hopkins University Press, Baltimore,

1954; "The beginnings of Operations Research: 1934–1941," J. F. McCloskey, *Operations Research*, 35, 1, 1987, 143–152; "British Operational Research in World War II," J. F. McCloskey, *Operations Research*, 35, 3, 1987, 453–470; "Air defence of Great Britain, 1920–1940: An operational research perspective," M. Kirby, R. Capey, *Journal of the Operational Research Society*, 48, 6, 1997, 555–568; *Operational Research in War and Peace*, M. W. Kirby, World Scientific, London, 2003]

The three rings of Blackett's Circus:

During World War II, Blackett and his Circus analysts established OR groups in the Anti-Aircraft (Ack-Ack) Command and in the Coastal Command for the Royal Navy, and were the precursors for similar groups in the British Army. Blackett's analysis for the Coastal Command proved to be of great significance. The analysis led to the use of shallower settings for depth charges and thus increased the sinking and damage of German U-boats. Blackett had been a naval officer, and at the age of 17, saw action in World War I, the naval battle off the Falkland Islands, and later at the battle of Jutland. Blackett won the 1948 Nobel prize in physics for his development of the Wilson cloud chamber method, and his discoveries therewith in the fields of nuclear physics and cosmic radiation.

(©Nobel Foundation)

Patrick M. S. Blackett
1897–1974

1941 Transportation problem

The first statement of the classical transportation problem (the shipping of goods from supply origins to demand destinations at minimum cost) is due to Frank L. Hitchcock in a 1941 paper in which he also sketched out a solution procedure. During World War II, the economist Tjalling C. Koopmans, while working for the British–American Combined Shipping Board, independently investigated and solved the same problem, thus the problem is referred to as the Hitchcock–Koopmans transportation problem. The formal statement of the transportation problem, its theory and computational resolution based on the simplex method, are due to George B. Dantzig. Early investigations of the transportation problem include that of A. N. Tolstoĭ in 1930 (see Schrijver, 2002) and that of Leonid V. Kantorovich (1942). ["Distribution of a product from several sources to numerous localities," F. L. Hitchcock, *Journal of Mathematical Physics*, 3, 1941, 224–230; "On the translocation of masses," L. V. Kantorovich, *Doklady Akad. Nauk SSSR*, 37, 7–8, 1942, 199–201, translated in *Management Science*, 5, 1, 1958, 1–4; "A model of transportation," T. C. Koopmans, S. Reiter, Chapter XIV (pp. 222–259) in *Activity Analysis of Production and Allocation*, T. C. Koopmans, editor, John Wiley & Sons, New York, 1951; "Application of the simplex method to a tranportation problem," G. B. Dantzig, Chapter XXIII (pp. 359–373) in *Activity Analysis of Production and Allocation*, T. C. Koopmans, editor, John Wiley & Sons, New York, 1951; *Linear Programming and Extensions*, G. B. Dantzig,

Princeton, 1963; "On the history of the transportation problem and maximum flow problems," A. Schrijver, *Mathematical Programming*, B, 91, 2002, 437–445]

1942 U.K. naval operational research

In December 1941, when Patrick M. S. Blackett was consulted about the formation of an OR section for the Admiralty, he wrote a memorandum entitled "Scientists at the operational level" which proved to be influential on both sides of the Atlantic; it provided an impetus for the formation of the U.S. Navy Antisubmarine Warfare Operations Research Group (ASWORG). In January 1942, Blackett moved to the British Admiralty to establish an OR group. The group scored an important success when it addressed the optimal size of a merchant convoy in terms of minimizing losses from submarine action and escort requirements. The Admiralty study clearly indicated that larger convoys are more effective and the implementation of this recommendation reduced losses substantially. Körner (1996) gives an intriguing account of Blackett's work and the mathematics and history of the military OR studies conducted by his group. ["A history of Operations Research," F. N. Trefethen, pp. 3–35 in *Operations Research for Management*, J. F. McCloskey, F. N. Trefethen, editors, The John Hopkins University Press, Baltimore, 1954; *Studies of War, Nuclear and Conventional*, P. M. S. Blackett, Hill and Wang, New York, 1962; *O.R. in World War 2: Operational Research against the U-boat*, C. H. Waddington, Paul Elek, London, 1973; "The beginnings of Operations Research: 1934–1941," J. F. McCloskey, *Operations Research*, 35, 1, 1987, 143–152; "British Operational Research in World War II," J. F. McCloskey, *Operations Research*, 35, 3, 1987, 453–470; *The Pleasures of Counting*, T. W. Körner, Cambridge University Press, Cambridge, U.K., 1996; *Operational Research in War and Peace*, M. W. Kirby, World Scientific, London, 2003]

1942 U.S. Navy Antisubmarine Warfare Operations Research Group (ASWORG)

ASWORG was the first civilian-staffed organization engaged in military OR in the U.S. It was organized for the Navy by the physicist Philip M. Morse early in World War II. It started with 15 civilian scientists assigned to the Office of Chief of Naval Operations, Admiral Ernest J. King. By the end of the war, there were almost 100 analysts employed by the more general problem-solving Operations Research Group (ORG), with ASWORG one of its subgroups. After the war, ORG was renamed the Navy's Operations Evaluation Group (OEG). In 1962, OEG was merged into the newly formed Center for Naval Analyses. Search theory had its beginnings in ASWORG with George E. Kimball and Bernard O. Koopman being the prime movers. Also a member of ASWORG was William Shockley who received the 1956 Nobel prize (joint with John Bardeen and Walter H. Brattain) for research on semiconductors and discovery of the transistor effect. [*Methods of Operations Research*, P. M. Morse, G. E. Kimball, John Wiley & Sons, New York, 1951; "Edison and Operations Research," W. F. Whitmore, *Journal of the Operations Research Society of America*, 1, 2, 1952, 83–85; *In at the Beginnings: A Physicist's Life*, Philip M. Morse, MIT Press, Cambridge, 1977; "Center for Naval Analyses," C. M. Harris, pp. 79–83 in *Encyclopedia of Operations Research and Management Science*, 2nd edition, S. I. Gass, C. M. Harris, editors, Kluwer Academic Publishers, Boston, 2001; "Search

Theory," L. D. Stone, pp. 742–745 in *Encyclopedia of Operations Research and Management Science*, 2nd edition, S. I. Gass, C. M. Harris, editors, Kluwer Academic Publishers, Boston, 2001]

The light bulb and beyond:

A precursor to ASWORG's activities was Thomas Edison's 1917 World War I statistical review of enemy submarine activity and U.S. and British shipping routes and procedures. Edison proposed a number of measures to the British Admiralty for reducing sinkings that were never acted upon.

Philip M. Morse, Robert H. Rinehart, Jacinto Steinhardt, Bernard O. Koopman, and George E. Kimball, all members of ASWORG, were presidents of the Operations Research Society of America (founded in 1952) in 1952, 1953, 1954, 1957, 1964, respectively.

1942 U.S. Air Force operations research

In October 1942, at the height of World War II, the first contingent of U.S. operations research analysts arrived in England to work with the Air Force's Eighth Bomber Command (later designated the Eighth Air Force). They were: James Alexander, mathematician from the Institute of Advanced Study at Princeton; Leslie H. Arps and John M. Harlan, lawyers from the New York law firm of Root, Ballantine, Harlan, Bushby and Palmer; H. P. Robertson, a physicist from Princeton University; W. Norris Tuttle, director of research at General Radio Company; William J. Youden, biochemist and statistician, and Boyce Thompson from the Plant Research Institute. Harlan was chief of this newly formed Operations Research Section which operated directly under the Chief of Staff. The analysts were instructed to first concentrate on improving bombing accuracy. Based on their quantitative studies of past bombing raids, they proposed that the best bombardier be in the lead airplane so as to aim the whole pattern of bombs, that all bombs be dropped in a salvo, and the aircraft fly in a tight precision formation, thus greatly reducing the bombing pattern dispersion. Based on the results of the OR analyses, there was at least a 1000 percent increase in bombs on target. ["Operations analysis in the United States Air Force," L. A. Brothers, *Operations Research*, 2, 1, 1954, 1–16; *Operations Analysis in the U.S. Army Eighth Air Force in World War II*, C. W. McArthur, History of Mathematics, Vol. 4, American Mathematical Society, Providence, 1990]

Beyond OR:

John H. Harlan was appointed by President Eisenhower to the United States Supreme Court (1955–1971). William J. Youden joined the Applied Mathematics Division of the National Bureau of Standards war and is noted for his work on experimental design.

1942 Search theory

Search theory deals with the problem of a searcher who wishes to find a target in an efficient manner. It had its beginnings in World War II when staff of the U.S. Navy's Antisubmarine Warfare Operations Research Group (ASWORG) investigated the German submarine threat in the Atlantic. The originally classified report, "Search and Screening" by Bernard O. Koopman was the first publication to describe a probabilistic based approach to the optimal allocation of search effort. ["Search and Screening," B. O. Koopman, Operations Evaluation Group Report No. 56, Center for Naval Analyses, Alexandria, 1946; "New mathematical methods of Operations Research," B. O. Koopman, *The Journal of the Operations Research Society of America*, 1, 1, 1952, 3–9; *Theory of Optimal Search*, L. D. Stone, Academic Press, New York, 1975; "Search Theory," P. M. Morse, pp. 485–544 in *Handbook of Operations Research*, J. J. Moder, S. E. Elmahraby, editors, Van Nostrand Reinhold, New York, 1978; *Search and Screening: General Principles and Historical Applications*, B. O. Koopman, Pergamon Press, New York, 1980; "Search Theory," L. D. Stone, pp. 742–745 in *Encyclopedia of Operations Research and Management Science*, 2nd edition, S. I. Gass, C. M. Harris, editors, Kluwer Academic Publishers, Boston, 2001]

The first among many:

Bernard O. Koopman was a founding member of ORSA and its sixth president in 1957. The Military Application Society awards the Koopman Prize each year for the outstanding publication in military operations research of the previous year. From 1959–61 he was the OR liaison between the U.S. Department of Defense, the U.K. military establishments, and NATO. His paper "New Mathematical Methods of Operations Research" was presented at the founding meeting of ORSA, May 27, 1952, and was the first technical paper published in *The Journal of the Operations Research Society of America*, Koopman (1952).

Bernard O. Koopman
1900–1981

1943 Neural networks

Warren S. McCulloch and Walter H. Pitts introduced the notion of a neural net as an abstraction of the physiological properties of nervous systems. They opened their seminal paper on the subject with the statement: "Because of the all-or-nothing character of nervous activity, neural events and the relations among them can be treated by propositional logic." Starting with the fact that each neuron reacts to excitation by either releasing a signal or failing to do so, McCulloch and Pitts showed that neural networks were capable of performing certain logical operations. In fact, the McCulloch–Pitts network could duplicate certain capabilities of a Turing machine, in other words, neural networks could compute. John von Neumann adapted the logical notation of McCulloch–Pitts in his logical description of the

EDVAC computer in 1945. The notion that neural networks could also learn was advanced by the McGill University physiologist Donald O. Hebb in 1949. Later, neural networks were more broadly defined as architectures based on connections among a set of neuron-like nodes and a variety of different architectures were proposed and studied. ["A logical calculus of the ideas immanent in nervous activity," W. S. McCulloch, W. H. Pitts, *Bulletin of Mathematical Biophysics*, 5, 1943, 115–133, reprinted in *Embodiments of Mind*, W. S. McCulloch, MIT Press, Cambridge, 1989, 19–39; *The Organization of Behavior*, D. O. Hebb, John Wiley & Sons, New York, 1949; *The Computer from Pascal to von Neumann*, H. M. Goldstine, Princeton University Press, Princeton, 1972; *AI: The Tumultuous History of the Search for Artificial Intelligence*, D. Crevier, Basic Books, New York, 1993]

1944 Exponential smoothing

As conceived by Robert G. Brown, exponential smoothing is "the name for a very special kind of weighted moving average. The new estimate of the average is updated periodically as the weighted sum of demand in the period since the last review and the old average. Thus it is not necessary to keep any record of past demand, the data processing becomes more economical." Brown first formalized the method around 1944 with continuous variables in the analysis of a fire control device. In the 1950s, he adapted the method to discrete variables, and featured it prominently in his 1959 text. In this work, Brown also proposed the use of Mean Absolute Deviation as a measure of dispersion for use in the statistical inventory control. Later, Brown extended exponential smoothing to handle a secular trend. [*Statistical Forecasting for Inventory Control*, R. G. Brown, McGraw-Hill, New York, 1959; *Smoothing, Forecasting, and Prediction*, R. G. Brown, Prentice-Hall, Englewood Cliffs, 1963; "Exponential Smoothing," R. G. Brown, pp. 275–277 in *Encyclopedia of Operations Research and Management Science*, 2nd edition, S. I. Gass, C. M. Harris, editors, Kluwer Academic Publishers, Boston, 2001]

1944 Modern utility theory

Utility theory is the systematic study and quantitative representation of preference structures. The idea of utility goes back to Daniel Bernouilli (1738), with the term popularized by Jeremy Bentham in 1789. The evolution of the concept can be found in Savage (1954) and in the readings collected by Page (1968). John von Neumann and Oskar Morgenstern provided the first axiomatic treatment of utility in the second edition of their classic work. [*Theory of Games and Economic Behavior*, J. von Neumann, O. Morgenstern, 2nd edition, Princeton University Press, Princeton, 1947; *The Foundations of Statistics*, L. J. Savage, John Wiley & Sons, New York, 1954; *Utility Theory: A Book of Readings*, A. N. Page, editor, John Wiley & Sons, New York, 1968; "What were von Neumann and Morgenstern trying to accomplish?" P. Mirowski, pp. 113–147 in *Toward a History of Game Theory*, E. R. Weintraub, editor, Duke University Press, Durham, 1992]

1944 *Theory of Games and Economic Behavior*, John von Neumann, Oskar Morgenstern, Princeton University Press, Princeton

This seminal book set forth the basic concepts of games of strategy and their application to economic and social theory. The revised 1947 edition is considered the standard reference; it includes, as an appendix, the authors' first statement of an axiomatic derivation of numerical utility theory.

The book:

THEORY OF GAMES AND ECONOMIC BEHAVIOR

By JOHN VON NEUMANN, and OSKAR MORGENSTERN

To Saul I. Gass with esteem Oskar Morgenstern

PRINCETON
PRINCETON UNIVERSITY PRESS
1947

John von Neumann 1903–1957

(Courtesy The American Economic Review) Oskar Morgenstern 1902–1976

1945 Project RAND

At the close of World War II, there was a need to have the services of scientists who could work on military planning and related U.S. government problems. To this end, the government established Project RAND (**R**esearch **an**d **D**evelopment) in December 1945 under contract to the Douglas Aircraft Company. ["RAND Corporation," G. H. Fisher, W. E. Walker, pp. 690–695 in *Encyclopedia of Operations Research and Management Science*, 2nd edition, S. I. Gass, C. M. Harris, editors, Kluwer Academic Publishers, Boston, 2001]

1945 U.S. Navy Operations Evaluation Group (OEG)

Due to decreased enemy submarine activity and the need for a broader application of OR to Navy problems, the Navy's first operations research group, the U.S. Antisubmarine Warfare Operations Research Group (ASWORG), was renamed the Operations Research Group (ORG) and assigned to Headquarters of the Atlantic Fleet. After the war, Admiral Ernest J. King wrote a letter to the Secretary of the Navy James Forrestal requesting that the Operations Research Group continue into peacetime at about a quarter of its wartime size. In November 1945, under a contract to the Massachusetts Institute of Technology (MIT), ORG was reconstituted as the Operations Evaluation Group (OEG), with Jacinto Steinhardt its first director. OEG published a number of important reports on naval operations, some reflecting wartime work that was originally classified, including such OR classics as Philip M. Morse's and George E. Kimball's *Methods of Operations Research* and Bernard Koopman's *Search and Screening*. Planning assumed an increasingly important role in defense and, prior to the Korean war, OEG slowly built up its staff to reach 40 by 1950. By the end of that war, it had grown to 60 research staff members. In 1962, OEG and the Institute for Naval Studies were merged in a new entity named the Center for Naval Analyses (CNA). ["Center for Naval Analyses," C. M. Harris, pp. 79–83 in *Encyclopedia of Operations Research and Management Science*, 2nd edition, S. I. Gass, C. M. Harris, editors, Kluwer Academic Publishers, Boston, 2001]

CNA = f(ASWORG, ORG, OEG):

Jacinto Steinhardt was a founding member of ORSA and served as its third president in 1957. He was the first and only Director of the Operations Evaluation Group. The Steinhardt Prize of the Military Applications Society of INFORMS is awarded periodically to a person whose life's work made outstanding contributions to Military Operations Research.

Jacinto (Jay) Steinhardt
1906–1985

1945 The diet problem

The economist, George Stigler, posed and analyzed the following problem: For a moderately active man (economist) weighing 154 pounds, how much of each of 77 foods should be eaten on a daily basis so that the man's intake of nine nutrients (including calories) will be at least equal to the recommended dietary allowances (RDAs) suggested by the National Research Council in 1943, with the cost of the diet being minimal? Stigler stated this optimizing problem in terms of a (9×77) set of simultaneous linear inequalities. As this was prior to George B. Dantzig's formalization of linear programming, Stigler had no exact procedure for finding the minimal cost solution. He astutely managed to find a nonoptimal solution that cost \$39.93. In 1947, Dantzig formulated Stigler's problem as a linear program and used it to test whether the simplex method would work well for a rather "large-scale" problem. A solution, using desk calculators and requiring 120 person-days of effort, was found with the optimal cost of \$39.69. Stigler received the 1982 Noble prize in economics for his seminal studies of industrial structures, functioning of markets and causes and effects of public regulation. ["The cost of subsistence," G. Stigler, *Journal of Farm Economics*, 27, 1945, 303–314; *Linear Programming and Extensions*, G. B. Dantzig, Princeton, 1963; "Stigler's diet problem revisited," S. I. Gass, S. Garille, *Operations Research*, 49, 1, 2001, 1–13]

Popeye was right:

Stigler's solution to his diet problem consisted of cabbage, dried navy beans, evaporated milk, spinach, and wheat flour. The linear-programming optimal solution used beef liver, cabbage, dried navy beans, spinach, and wheat flour.

George Stigler
1911–1991

1946 The digital computer

There is much prehistory to the development of the digital computer, starting with Charles Babbage and his analytical engine. The reader is referred to the book by Herman Goldstine (1972) for a rather detailed discussion of the events and people that led to the development of the digital computer. The year 1946 saw the debut of what is considered the first modern general-purpose digital computer, the ENIAC (Electronic Numerical Integrator and Computer). The field of operations research would not have expanded as it did in the late 1940s and 1950s without the synergistic influence of the computer, e.g., the development and use of Monte Carlo and discrete simulation, and the solution of linear-programming problems in government and industry by the simplex method. [*The Computer from Pascal to von Neumann*, H. M. Goldstine, Princeton University Press, Princeton, 1972; *A Computer Perspective*, G. Fleck, editor, Harvard University Press, Cambridge, 1973; *ENIAC: The Triumphs and Tragedies of the World's First Computer*, S. McCartney, Berkley Books, New York, 1999; *From* 0 *to* 1: *An Authoritative History of Modern Computing*, A. Akera, F. Nebeker, editors, Oxford University Press, New York, 2002]

1946 Monte Carlo simulation

The Monte Carlo method was the idea of the mathematician and theoretical physicist Stanislaw Ulam, who thought of it while playing solitaire during an illness in 1946. It was first announced by Ulam and John von Neumann in a short abstract submitted to the American Mathematical Society on September 3, 1947. They stated that this procedure was "... analogous to the playing of a series of 'solitaire' card games and is performed on a computing machine. It requires ... 'random' numbers with a given distribution." The roots of the method reside in von Neumann's use of the computer to obtain results for complex physics problems. An important example was the use of numerical techniques to study the hydrodynamics of the implosion necessary to trigger a nuclear detonation. In 1945, von Neumann invited Stanley Frankel and Nicolas Metropolis to tackle the difference equations associated with the thermonuclear weapon (the superbomb) on the new ENIAC. Ulam was present at the April 1946 superbomb meeting where Metropolis and Frankel presented their ENIAC results; the need for more efficient ways of obtaining the results was clear. The term Monte Carlo was coined by Metropolis and appeared in his joint 1949 paper with Ulam. ["On combination of stochastic and deterministic processes: Preliminary reports," S. M. Ulam, J. von Neumann [abstract], *Bulletin of the American Mathematical Society*, 53, 1947, 1120; "The Monte Carlo Method," N. Metropolis, S. M. Ulam, *Journal of the American Statistical Association*, B, 44, 1949, 335–341; *From Cardinals to Chaos: Reflections on the Life and Legacy of Stanislaw Ulam*, N. G. Cooper, editor, Cambridge University Press, New York, 1989; "The transformation of numerical analysis by the computer: An example from the work of John von Neumann," W. Aspray, pp. 307–322 in *The History of Modern Mathematics, Vol. II: Institutions and Applications*, D. E. Rowe, J. McCleary, editors, Academic Press, Boston, 1989; "Statistical methods in neutron diffusion," J. von Neumann, R. D. Richtmyer, Los Alamos Scientific Laboratory Report LAMS-551, April 1947, pp. 16–36 in *Analogies Between Analogies: The Mathematical Reports of S. M. Ulam and his Los Alamos Collaborators*, A. R. Bednarek, F. Ulam, editors, University of California Press, Berkeley, 1990; *Image and Logic*, P. Gallison, University of Chicago Press, Chicago, 1997]

1946 ***Mathematical Methods of Statistics***, Harald Cramér, Princeton University Press, Princeton

The purpose of this book was to join the modern mathematical theory of probability with statistical science, as developed by Ronald A. Fisher and his British and American contemporaries. Harald Cramér was known for the clarity of his lectures and writings. The roots of this book go back to his classroom lectures of the 1930s, but the text was mainly written during 1942–1944. The first two parts of the book develop the foundations, while the third part, which comprises over 40% of the book, is devoted to statistical inference. In the words of Leadbetter (2001), this book "provided a wonderfully timely and lucid account of a hitherto hodgepodge of often mysterious statistical procedures, now organized as a coherent mathematical discipline .... It has had immense influence on generations of statisticians and especially ... in encouraging young mathematicians to enter and find a mathematically satisfying career in statistics." ["Harald Cramér," M. R. Leadbetter, pp. 439–443 in *Statisticians of the Centuries*, G. C. Heyde, E. Seneta, editors, Springer-Verlag, New York, 2001]

1946 ***Methods of Operations Research***, Philip M. Morse, George E. Kimball, (Classified), Operations Evaluation Group, OEG Report 54, U.S. Department of the Navy, Washington, DC (Unclassified version, MIT Press and John Wiley & Sons, New York, 1951; Dover reprint 2003)

The unclassified version introduced the basic concepts of OR to U.S. industrial, business, and nonmilitary governmental executives, as well as to the academic research community. It invoked and popularized an early definition of OR: "Operations Research is a scientific method of providing executive departments with a quantitative basis for decisions regarding the operations under their control."

Prime movers:

After the war, Morse returned to MIT. He was appointed Chairman of the Committee on Operations Research (1952), organized the first summer seminars in OR, and formed and directed the cross-campus MIT Operations Research Center (1955). He had a distinguished career in physics and OR, and in the administration of major scientific endeavors. Morse was a founding member and first president of the Operations Research Society of America (1952).

Philip M. Morse
1903–1985

Kimball returned to the Chemistry Department of Columbia University after World War II. He was an early visionary with respect to applying and extending the wartime developments of OR into the business and industrial domains. In 1956, he joined the Cambridge consulting firm of Arthur D. Little as its first Science Advisor, and became Vice President in 1961. Kimball was a founding member of the Operations Research Society of America and served as its president in 1964.

George E. Kimball
1906–1967

4

Expansion of operations research from 1947 to 1950

1947 Project SCOOP (U.S. Air Force Scientific Computation of Optimal Programs)

Project SCOOP (Scientific Computation of Optimal Programs) was a Pentagon-based U.S. Air Force research group formed in June 1947. It was officially designated Project SCOOP in October 1948 and disbanded in 1955. It was headed by the economist Marshall K. Wood, with George B. Dantzig chief mathematician. The main objective of Project SCOOP was to develop more suitable answers to the problem of programming Air Force requirements, for example, determining the time-phased requirements of materials to support a war plan. It was at Project SCOOP where Dantzig first stated the mathematical form of the general linear program and, along with Wood, established the related mathematical and economic theories of program planning – the selection of competing, interdependent activities – so as to determine a program that best meets objectives without exceeding resource limitations. The mathematical structure of the linear-programming problem is a generalization of Leontief's static interindustry model in that it explicitly considers the dynamic aspects of program planning. It was also at Project SCOOP where Dantzig invented the simplex method for solving such problems and where both the linear-programming model and the simplex method were tested and proven. Dantzig's seminal work on the simplex algorithm, the simplex transportation algorithm, and the relationship between linear programming and zero-sum two-person games was first (formally) published in *Activity Analysis of Production and Allocation*, T. C. Koopmans, editor, John Wiley & Sons, New York, 1951. ["Programming of inter-dependent activities I, General discussion," M. Wood, G. B. Dantzig, *Econometrica*, 17, 3–4, 1949, 193–199; "Programming of inter-dependent activities II, Mathematical Model," G. B. Dantzig, *Econometrica*, 17, 3–4, 1949, 200–211; "The mathematical computation branch: Origins, functions, and facilities," DCS/Comptroller, U.S. Air Force, Washington, DC, 1953; "Concepts, origins, and use of linear programming," G. B. Dantzig, Report P-980, The RAND Corporation, Santa Monica, 1957; *Linear Programming and Extensions*, G. B. Dantzig, Princeton University Press,

Princeton, 1963; "Reminiscences about the origins of linear programming," G. B. Dantzig, *Operations Research Letters*, 1, 2, 1982; "Linear programming," G. B. Dantzig, *Operations Research*, 50, 1, 2002, 42–47; "The first linear programming shoppe," S. I. Gass, *Operations Research*, 50, 1, 2002, 61–68]

SCOOP crystal ball:

From Dantzig and Wood (1949): "To compute programs rapidly with such a mathematical model (linear programming), it is proposed that all necessary information and instructions be systematically classified and stored on magnetized tapes in the 'memory' of a large scale digital electronic computer. It will then be possible, we believe, through the use of mathematical techniques now being developed, to determine the program which will maximize the accomplishment of our objectives within those stated resource limitations."

From DCS/Comptroller, U.S. Air Force (1953): "The work of the Planning Research Division (Project SCOOP) with the three models (rectangular optimization model, square model of linear equations, triangular square model of linear equations) has given considerable impetus to the current interest in models of linear equations – or 'linear models,' as they are becoming to be known. Following a term contributed by the Division, this field is widely known among mathematicians as 'linear programming,' although 'activity analysis' is gaining favor. It is hard to say whether more attention is directed towards rectangular models or square models, but it is clear that many mathematicians view the rectangular model (the linear-programming model) as one with a great future. In a world where more and more attention is certain to be given to the efficient allocation of money and resources – in various situations from the national level right down to the plant or process level – the rectangular model is naturally exciting."

1947 The linear-programming problem

Programming problems are concerned with the efficient use or allocation of limited resources to meet desired objectives. Typical examples are refinery operations that transform crude oil into different fuels, transportation of material from many sources to many destinations, and the production of goods to meet demand. A linear-programming problem can be stated mathematically as follows: Minimize (or Maximize) cx, subject to $Ax = b$, $x \geqslant 0$, where c is a $(1 \times n)$ row vector, x is a $(n \times 1)$ column vector, A is an $(m \times n)$ matrix, and b is a $(m \times 1)$ column vector. First stated in this form by George B. Dantzig, it is an amazing fact that literally thousands of decision (programming) problems from business, industry, government and the military can be stated (or approximated) as linear-programming problems. Although there were some precursor attempts at stating such problems in mathematical terms, notably by the Russian mathematician Leonid V. Kantorovich in 1939, Dantzig's general formulation, combined with his method of solution, the simplex method, revolutionized decision making in the second half of the twentieth century. The name "linear programming" was suggested to Dantzig by the economist Tjalling C. Koopmans. Both Kantorovich and Koopmans were awarded the 1975 Nobel prize in economics for their contributions to the theory of optimum allocation of resources. ["On the translocation of masses," L. V. Kantorovich, *Doklady Akad. Nauk SSSR*, 37, 7–8, 1942, 199–201, translated in *Management Science*, 5, 1, 1958, 1–4; "Mathematical methods of organization and planning production," L. V. Kantorovich, Publication House of the Leningrad State University, 1939, translated in *Management Science*, 6, 4, 1960, 366–422; *Linear Programming and Extensions*, G. B. Dantzig, Princeton University Press, Princeton, 1963; "The discovery of linear programming," R. Dorfman, *Annals of the History of Computing*, 6, 3, 1984, 283–295; "My journey into science (Supposed report to the Moscow Mathematical Society)," posthumous report of L. V. Kantorovich, prepared by V. L. Kantorovich, *Russian Mathematics Surveys*, 42, 2, 1987, also reprinted in *Functional Analysis, Optimization, and Mathematical Economics: A Collection of Papers Dedicated to the Memory of Leonid Vital'evich Kantorovich*, L. J. Leifman, editor, Oxford University Press, New York, 1990, 8–45; "Comments on the history of linear programming," S. I. Gass, *Annals of the History of Computing*, 11, 2, 1989, 147–151; "L. V. Kantorovich: The price implications of optimal planning," R. Gardner, *Journal of Economic Literature*, 28, June 1990, 638–648; "Mathematical programming: Journal, society, recollections," M. L. Balinski, pp. 5–18 in *History of Mathematical Programming*, J. K. Lenstra, A. H. G. Rinnooy Kan, A. Schrijver, editors, North-Holland, Amsterdam, 1991; "Linear programming," G. B. Dantzig, *Operations Research*, 50, 1, 2002, 42–47]

Linear programming, the Nobel prize, and Marxist economics:

Most people familiar with the origins and development of linear programming were amazed and disappointed that Dantzig did not receive the Nobel prize along with Koopmans and Kantorovich (a Nobel prize can be shared by up to three recipients). According to Michel L. Balinski (1991), Koopmans was profoundly distressed that Dantzig had not shared in the prize. Koopmans gave a gift of $40,000 to the International Institute for Applied Systems Analysis (IIASA) in Laxenburg, Austria, the amount equal to his share of what Dantzig would have received. All three principals had worked and met at various times at IIASA. In a conversation we had with Koopmans shortly after the award, he told of his displeasure with the Nobel selection and how he had earlier written to Kantorovich suggesting that they both refuse the prize, certainly a most difficult decision for both, but especially so for Kantorovich. His work in this area received little recognition in the Soviet Union when it was first developed. As Kantorovich noted (in a posthumous publication, 1987): "In the spring of 1939 I gave some more reports – at the Polytechnic Institute and the House of Scientists, but several times met with the objection that the work used mathematical methods, and in the West the mathematical school in economics was an anti-Marxist school and mathematics in economics was a means for apologists of capitalism." Dantzig (1963) notes: "Kantorovich should be credited with being the first to recognize that certain important broad classes of production problems had well-defined mathematical structures which, he believed, were amenable to practical numerical evaluation and could be numerically solved."

Tjalling C. Koopmans 1901–1981 George B. Dantzig Leonid V. Kantorovich 1912–1986

1947 Simplex method

The (primal) simplex algorithm was invented by George B. Dantzig as a solution procedure for solving linear-programming (LP) problems. It has been used to solve a wide-variety of such problems most efficiently on all types of digital computers, beginning with the early (very slow and cumbersome) varieties of the late 1940s and early 1950s to the high-speed computers of the 21st Century. The algorithm starts with a basic feasible solution and then searches a finite sequence of other basic feasible solutions until one is found that also satisfies optimality conditions. Other methods for solving LP problems have since been developed, in particular interior-point methods, but the simplex method is the workhorse of LP. The simplex method was picked as one of the 20th Century's top ten algorithms. [*Linear Programming and Extensions*, G. B. Dantzig, Princeton University Press, Princeton, 1963; "The top ten algorithms of the century," in supplement to *Computing in Science and Engineering*, 1, 6, IEEE, 2000; "Linear programming," G. B. Dantzig, *Operations Research*, 50, 1, 2002, 42–47]

1947 The acceptance-rejection method for generating random variates

Given a source of randomly generated numbers with the uniform distribution on [0, 1], how does one generate a random variate X with a known probability distribution function $F(x)$? Two well-known methods for this are the inverse transform method (which inverts the function F) and the acceptance-rejection method. The latter uses pairs of independent uniform random numbers Y and U and accepts the value of Y when it satisfies $U \leqslant f(Y)$. As stated here, this procedure assumes that the density function $f(x)$ lies between 0 and 1. This method was proposed by John von Neumann in a letter dated March 21, 1947 to Stanislaw Ulam. The letter also described the inversion method, which Ulam had already thought of. ["Stan Ulam, John von Neumann, and the Monte Carlo Method," R. Eckhardt, pp. 131–137 in *From Cardinals to Chaos: Reflections on the Life and Legacy of Stanislaw Ulam*, N. G. Cooper, editor, Cambridge University Press, New York, 1989]

1947 The Association for Computing Machinery (ACM) founded

The Association for Computing Machinery is an international scientific and educational organization dedicated to advancing the art, science, engineering, and application of information technology. Its first president was John H. Curtis.

1947 *Sequential Analysis*, Abraham Wald, John Wiley & Sons, New York (Dover reprint 1973)

This book describes Abraham Wald's seminal work on sequential tests of statistical hypotheses. According to Wald, the resulting "sequential probability ratio test frequently results in a savings of about 50 per cent in the number of observations over the most efficient test procedure based on a fixed number of observations." The problem arose within the Statistical Research Group (SRG) led by W. Allen Wallis at Columbia University, which operated under a contract with the Office of Scientific Research and Development during World War II. Captain G. L. Schuyler of the Navy's Bureau of Ordnance asked Wallis if there was a sampling plan that could terminate a statistical experiment earlier than planned *ex ante*. Wallis discussed the problem with Milton Friedman (1976 Nobel prize in economics) and they both realized the importance of this research question and its potential to reduce the sample size required. Wallis and Friedman approached Wald, who developed the theory in 1943 and published the results in his classic 1944 paper. Prior to this work, Wald was a pioneer in operations analysis and had written a paper on how to measure the vulnerability of aircraft from damage data collected from returning aircraft. ["On cumulative sums of random variables," A. Wald, *The Annals of Mathematical Statistics*, 15, 1944, 283–296; "The Statistical Research Group, 1942–1945," W. Allen Wallis, *Journal of the American Statistical Association*, 75, 370, 1980, 320–330; "Mathematicians at War: Warren Weaver and the applied mathematics panel, 1942–1945," L. Owens, pp. 287–

Abraham Wald
1902–1950

305 in *The History of Modern Mathematics, Vol. II: Institutions and Applications*, D. E. Rowe, J. McCleary, editors, Academic Press, Boston, 1989; "A conversation with Herbert Solomon," P. Switzer, *Statistical Science*, 7, 3, 1992, 388–401]

1947 The definition of OR

The paper by Charles Kittel (1947) is one of the first papers that brought the ideas of OR to the U.S. scientific community. As Kittel stated: "It is hoped that the publication of this paper will serve to stimulate the establishment of operations research groups in the United States for the advancement of peaceful objectives. This powerful new tool should find a place in government and industry." His article gives a concise statement of the origins of OR, and then describes World War II OR: thousand-plane raids, large merchant-vessel convoys, bombing of Japan, submarine wolf-packs, exchange rates (ratio of output to input), and effectiveness ratios (e.g., ships sunk/torpedoes fired). Most important, Kittel formulated the following definition of OR: "Operations Research is a scientific method for providing executive departments with a *quantitative basis for decisions.*" This definition was modified by Charles Goodeve (1948) to read: "Operational Research is a scientific method of providing executive departments with a quantitative basis for decisions regarding the operations under their control." The latter definition is the one popularized by Morse and Kimball (1951). Kittel, a physicist noted for his book *Introduction to Solid State Physics*, served as an operations analyst with the U.S. fleet from 1943–1945. He is a Professor Emeritus, University of California, Berkeley. ["The nature and development of Operations Research," C. Kittel, *Science*, 105, 2719, February 7, 1947, 105–153; "Operational research," C. Goodeve, *Nature*, 161, 4089, March 13, 1948, 377–384; *Methods of Operations Research*, P. M. Morse, G. E. Kimball, MIT Press and John Wiley & Sons, New York, 1951 (Dover reprint 2003); *An Annotated Bibliography on Operations Research*, Vera Riley, Operations Research Office, The Johns Hopkins University, Chevy Chase, 1953]

1948 OR in the industrial sector: British Iron and Steel Industry Research Association

The National Coal Board of Great Britain, formed in 1948, established a major OR activity in its Field Investigation Group headed by Berwyn Hugh Patrick Rivett. The major studies conducted by this group included colliery organization, communications and underground transport, coal distribution, and manpower analysis. The year 1948 also marked the formation of the British Iron and Steel Research Association (BISRA) with Sir Charles Goodeve as director. BISRA employed OR to address industry-wide problems and also helped the larger firms in the industry to establish their own OR groups. In particular, Stafford Beer headed a 70 plus group of OR professionals for United Steel. ["War and Peace: The first 25 years of OR in Great Britain," K. B. Haley, *Operations Research*, 50, 1, 2002, 82–88; "Stafford Beer, Obituary," J. Rosenhead, D. Martin, *OR Newsletter Quarterly*, October 2002, 16–17]

B. H. P. (Pat) Rivett

Charles Goodeve
1904–1980

Stafford Beer
1926–2002

1948 The RAND Corporation

In February 1948, Project RAND was converted into an independent nonprofit corporation. Over the years, RAND researchers contributed greatly to many OR areas: game theory, linear programming, dynamic programming, systems analysis, simulation, flows in networks, and the Delphi method. ["RAND Corporation," G. H. Fisher, W. E. Walker, pp. 690–695 in *Encyclopedia of Operations Research and Management Science*, 2nd edition, S. I. Gass, C. M. Harris, editors, Kluwer Academic Publishers, Boston, 2001]

1948 Johns Hopkins U.S. Army Operations Research Office (ORO)

During World War II, military operations research in the U.S. was conducted mainly by elements of the Army Air Corps and the Navy. It was only after the war that the U.S. Army formally established an OR activity, the General Research Office, under the management of the trustees of the Johns Hopkins University, located at Ft. McNair, Washington, DC. The name was soon changed to the Operations Research Office, and, in 1951, ORO moved to its long-term headquarters in Chevy Chase, MD. ORO's founding and only director was the geophysicist Ellis A. Johnson. ORO was disbanded in 1961, with its activities transferred to the newly formed Research Analysis Corporation, a Federal Contract Research Center. ["A history of Operations Research," F. N. Trefethen, pp. 3–35 in *Operations Research for Management*, J. F. McCloskey, F. N. Trefethen, editors, The John Hopkins University Press, Baltimore, 1954; "Ellis A. Johnson, 1906–1973," T. Page, G. D. Pettee, W. A. Wallace, *Operations Research*, 22, 6, 1974, 1139–1153; "Operations Research Office and Research Analysis Corporation," E. P. Visco, C. M. Harris, pp. 595–599 in *Encyclopedia of Operations Research and Management Science*, 2nd edition, S. I. Gass, C. M. Harris, editors, Kluwer Academic Publishers, Boston, 2001]

In at the beginning:

Ellis A. Johnson received a DSc degree in Electrical Engineering from MIT in 1934. He soon became involved in terrestrial magnetism, becoming head of the Mine Research Unit of the Naval Ordnance Laboratory. His unit developed a degaussing process for neutralizing the magnetic field of a ship's hull. He was in Pearl Harbor working on a ship degaussing range for the Pacific Fleet when the Japanese attacked on December 7, 1941. During the bombing, he boarded a minesweeper to help clear the harbor of mines (Page, Pettee, Wallace, 1974).

Ellis A. Johnson
1906–1973

1948 Operational Research Club of Great Britain

The OR Club was inaugurated in April 1948 in London with Sir Charles Goodeve as its chairman. The genesis of the Club was the need of a mutual support group for introducing OR into industry. To maintain the informal nature of the Club, membership was limited to 50. Hugh Miser describes the background of the OR Club as follows: "In April 1948 several scientists who had taken part in the successful development of operations research in England during World War II, and who had had occasional informal meetings to discuss their work, agreed among themselves to act as conveners of the Operations Research Club; J. A. Jukes became its first honorary secretary. Its purpose was to provide a continuing structure for these informal meetings, and six were held each year between September and May in the rooms of the Royal Society in London." In November 1953, the OR Club was restructured and expanded to form the Operational Research Society (U.K.). ["The history, nature, and use of Operations Research," H. J. Miser, pp. 3–24 in *Handbook of Operations Research, Foundations and Fundamentals*, Van Nostrand Reinhold, New York, 1978; "A history of OR in 2000 words," N. Cummings, *OR Newsletter*, April 2001, 20–23]

1949 Equipment replacement

Equipment replacement deals with determining the optimum point in time to replace a unit (economic life problem) and/or choosing the best equipment to replace a unit (equipment selection problem). George Terborgh was the first to develop a theory for equipment replacement. [*Dynamic Equipment Policy*, G. Terborgh, McGraw-Hill, New York, 1949; "Replacement Theory," B. V. Dean, pp. 327–362 in *Publications in Operations Research, No. 1*, R. L. Ackoff, editor, John Wiley & Sons, New York, 1961]

1949 Linear congruential random number generators

Computer-based random number generators that are most widely used are based on a method proposed by Derrick H. Lehmer in 1949. The method requires four integers:

$x_0 > 0$ (starting value), $a > 0$ (multiplier), $c \geqslant 0$ (increment), and $m > 0$ (modulus), with m greater than the other three in magnitude. These numbers are then related by a linear congruential number generator of the form, $x_{n+1} = (ax_n + c) \bmod m$, for $n \geqslant 0$. The resulting sequence consists of pseudorandom numbers. ["Mathematical methods in large-scale computing units," D. H. Lehmer, pp. 141–146 in *Proceedings of the Second Symposium on Large-Scale Digital Calculating Machinery,* Harvard University Press, Cambridge, 1951; "Various techniques used in connection with random digits," J. von Neumann, National Bureau of Standards, *Applied Mathematics Series*, 12, 1951, 36–38; "Random number generators," T. E. Hull, A. R. Dobell, *SIAM Review*, 4, 5, 1962, 230–254; *The Art of Computer Programming, Vol. 2: Seminumerical Algorithms,* D. E. Knuth, 2nd edition, Addison-Wesley, Reading, 1981]

True or pseudo:

Lehmer (1951) gave the following description of a pseudorandom sequence: "... a vague notion embodying the idea of a sequence in which each term is unpredictable to the uninitiated and whose digits pass a certain number of tests, traditional with statisticians and depending somewhat on the uses to which the sequence is to be put."

Von Neumann (1951) stated: "Any one who uses arithmetical methods to produce random numbers is, of course, in a state of sin."

Derrick H. Lehmer
1905–1991

1949 Cowles Commission conference

On June 20–24, 1949, at the University of Chicago, the Cowles Commission for Research in Economics sponsored a conference on "Activity Analysis of Production and Allocation." This conference is notable in that it was here that George B. Dantzig, Tjalling C. Koopmans, Harold W. Kuhn, Albert W. Tucker, and Marshall K. Wood, among others, presented papers that help to establish the theoretical and applied aspects of linear programming and its extensions. This conference is considered to be the 0th Mathematical Programming Symposium. [*Activity Analysis of Production and Allocation*, T. C. Koopmans, editor, John Wiley & Sons, New York, 1951]

1949 Cost effectiveness analysis

Cost effectiveness analysis is the process of using theory, data, and models to examine the relevant objectives of a problem and comparing the costs, benefits, and risks of alternative ways of achieving these objectives. After World War II, as the U.S. Armed Services began competing for missions, the approval of budgets for new systems had to be based on a sound procedure. Cost effectiveness emerged as the key criterion for such allocations. The analytical process of comparing alternative solutions was first called "weapon systems analysis" and later shortened to systems analysis. The first documented systems analysis was conducted by the RAND Corporation in comparing the B-52 to a turbo-prop bomber. The work of D. Novick at RAND in the 1950s led to a detailed process for cost analysis. [*A History of Cost Effectiveness*, E. S. Quade, U.S. Air Force Project, P-4557, The RAND Corporation, Santa Monica, 1971; "Beginnings of military cost analysis: 1950–1961," D. Novick, P-7425, The RAND Corporation, Santa Monica, 1988; "Cost analysis," S. J. Balut and T. R. Gulledge, pp. 152–155 in *Encyclopedia of Operations Research and Management Science*, 2nd edition, S. I. Gass, C. M. Harris, editors, Kluwer Academic Publishers, Boston, 2001; "Cost effectiveness analysis," N. K. Womer, pp. 155–158 in *Encyclopedia of Operations Research and Management Science*, 2nd edition, S. I. Gass, C. M. Harris, editors, Kluwer Academic Publishers, Boston, 2001]

1949 Arrow's impossibility theorem

Kenneth Arrow's impossibility theorem states that, in general, it is impossible to extend a set of individual preferences to a social preference ordering R that uses the information in the individual choices and satisfies certain highly desirable conditions. Two key conditions are: the Pareto principle (if everyone strictly prefers x to y, then x is preferred to y in R), and the irrelevance of independent alternatives (the choice between any two alternatives depends only on the preferences of individuals among those two alternatives). The roots of Arrow's thinking on this famous result are threefold: First, he had been exposed to the set-theoretic calculus of relations in a course taught by Alfred Tarski. Second, he had absorbed Harold Hotelling's ordinalist interpretation of utilities, and third, he had thought about aggregation of individual preferences as he was writing his dissertation in 1947. In 1949, Olaf Helmer, a philosopher and translator of Tarski's works, asked Arrow for a justification of the aggregation of individual utilities in a manner consistent with the ordinal concept. Arrow knew that majority voting would not aggregate appropriately, but assumed that an alternative scheme may work. After a few days, he realized that this may be an impossibility result. He presented the results in the 1949 meeting of the Econometric Society, with papers following in 1950 and 1951. Arrow was awarded the Nobel prize in 1972, joint with John R. Hicks, for their pioneering contributions to general economic equilibrium theory and welfare theory. [*Social Choice and Individual Values,* Kenneth J. Arrow, Cowles Commission Monograph 12, John Wiley & Sons, New York, 1951; *Arrow Impossibility Theorems,* J. Kelly, Academic Press, New York, 1978; "The origin of the impossibility theorem," K. J. Arrow, pp. 1–4 in *History of Mathematical Programming, A Collection of Personal Reminiscences*, J. K. Lenstra, A. H. G. Rinnooy Kan, A. Schrijver, editors, North-Holland, Amsterdam, 1991]

(©Nobel Foundation)
Kenneth Arrow

1949 Operations research at Arthur D. Little, Inc.

Several members of Arthur D. Little, Inc. (ADL), who had taken leave during World War II to enter government service, had worked in or had been exposed to military operations research. They convinced a senior officer of ADL, Raymond Stevens, to explore the use of OR in industry. Stevens asked Harry B. Wissman to build an OR group within ADL, one of the first nongovernmental OR consultancy groups. Wissman persuaded Sears, Roebuck & Co. to become a client for its services. Philip Morse and George Wadsworth of MIT were already consultants to ADL, as was George Kimball of Columbia University. Kimball became an ADL staff member in early 1950s. Wissman recruited John F. Magee to join the group; other members included John Lathrop, Sherman Kingsbury, Arthur Brown, Martin Ernst, and David Boodman. Kimball directed a project for the baby products division of Johnson & Johnson that utilized Magee's developments in production and inventory control. The logistics work at ADL led to articles published in the *Harvard Business Review*, forming the basis of Magee's 1958 book *Production Planning and Inventory Control*. ["Operations Research at Arthur D. Little, Inc.: The early years," John F. Magee, *Operations Research,* 50, 1, 2002, 149–153]

ORSA 2, TIMS 1:

John F. Magee was a founding member of ORSA and was president of both ORSA (1966) and TIMS (1971–1972). He joined Arthur D. Little in 1950 as a member of the Operations Research Group, eventually becoming its President, Chief Executive Officer, and Chairman of the Board.

John B. Lathrop was a founding member of ORSA, serving as its president in 1958. He was an OR analyst with the Navy's Operations Evaluation Group (OEG). Following OEG, he joined Arthur D. Little to work on manufacturing control, advertising, and quality control. He later joined Lockheed Aircraft, retiring as manager of systems analysis.

John F. Magee

John B. Lathrop
1910–1978

1949 *Extrapolation, Interpolation, and Smoothing of Stationary Time Series*, Norbert Wiener, John Wiley & Sons, New York

Written with a focus on engineering applications, this book became a cornerstone of furture work in prediction and optimal control. Its stated purpose was to unite the theory and practice of communications engineering and time series analysis. Most of the work reflects Wiener's own original contributions, which exploit the full force of Fourier methods to provide the methodological unity. Prediction and filtering problems are discussed for both single and multiple time series, and the theoretical links with harmonic analysis are pointed out. The book also discusses the notion of using optimal weights to predict moving averages of a time series which influenced the development of exponential smoothing and related time series forecasting methods.

1950 Statistical decision theory

In individual decision making under uncertainty, a choice must be made from a set of allowable actions $A_1, A_2, \ldots, A_m$, where the relative desirability of each action depends upon the prevailing state of nature. The decision-maker (DM) knows the possible states of nature $S_1, S_2, \ldots, S_n$ and the payoffs u_{ij} (utility or value) associated with each pair (A_i, S_j). It is generally assumed that the probability of each state occurring is not known with certainty. However, if an *a priori* probability distribution over the states of nature exists, or is assumed by the DM, then one can address decision-making under risk. This framework for decision-making was developed in the early 1950s and can be viewed as a precursor of modern decision analysis. ["Remarks on the rational selection of a decision function", H. Chernoff, Cowles Commission discussion paper (unpublished), Statistics, Nos. 326–326A, 1949, 422–443; *Statistical Decision Functions*, A. Wald, John Wiley & Sons, New York, 1950; "Optimal criteria for decision making under ignorance," L. Hurwicz, Cowles Commission discussion paper (unpublished), Statistics, No. 370, 1951; "The theory of statistical decision," L. J. Savage, *Journal of the American Statistical Association*, 46, 1951, 55–67; *The Foundations of Statistics*, L. J. Savage, John Wiley & Sons, New York, 1954; "Rational selection of decision functions," H. Chernoff, *Econometrica*, 22, 1954]

1950 First solution of the transportation problem on a computer

The simplex algorithm, adapted for solving the special structure of the transportation linear programming problem, was coded for the National Bureau of Standards SEAC digital computer under auspices of the USAF's Project SCOOP. A general simplex code was developed for the SEAC in 1951. [*Linear Programming and Extensions*, G. B. Dantzig, Princeton University Press, Princeton, 1963; "The first linear programming shoppe," S. I. Gass, *Operations Research*, 50, 1, 2002, 61–68]

1950 Post World War II quality control

W. Edwards Deming was a mathematical physicist in the Bureau of Chemistry and Soils, U.S. Department of Agriculture, where he was instrumental in introducing the ideas of modern statistical knowledge. His paper with R. T. Birge was influential in bringing the methods of Ronald A. Fisher, Jerzy Neyman, and Egon S. Pearson to American physical scientists, while his books on sampling and the design of business research disseminated the use of sampling beyond government. Deming's later fame resulted from his interest in quality control, which was influenced by the work of Walter A. Shewhart. From 1947–1950, Deming served as an advisor in sampling techniques to General MacArthur's supreme command in Tokyo, and as an advisor to the Japanese Union of Scientists and Engineers (JUSE). Deming's approach to statistical quality control was adopted widely by Japanese business and manufacturing and was a major force in the resurgence of the Japanese economy. In 1950, JUSE created the Deming Prize for excellence in quality. Deming's approach to total quality management (TQM) is put forth in his famous 14-point philosophy. ["On the statistical theory of errors," W. E. Deming, R. T. Birge, *Review of Modern Physics*, 6, 1934, 119–161; *Some Theory of Sampling*, W. E. Deming, John Wiley & Sons, New York, 1950 (Dover reprint 1966); *Statistical Design in Business Research*, W. E. Deming, John Wiley & Sons, New York, 1960; *Out of the Crisis*, W. E. Deming, MIT Press, Cambridge, 1986; *Statisticians of the Centuries*, G. C. Heyde, E. Seneta, editors, Springer-Verlag, New York, 2001; "Total quality management," J. S. Ramberg, pp. 836–842 in *Encyclopedia of Operations Research and Management Science*, 2nd edition, S. I. Gass, C. M. Harris, editors, Kluwer Academic Publishers, Boston, 2001]

Quotable Deming:

"If you can't describe what you are doing as a process, you don't know what you're doing."

"What we need to do is learn to work in the system, by which I mean that everybody, every team, every platform, every division, every component is there not for individual competitive profit or recognition, but for contribution to the system as a whole on a win-win basis."

"Experience teaches nothing without theory."

(The W. Edwards Deming Institute®)

W. Edwards Deming
1900–1993

1950 The prisoner's dilemma

A simply told story of a nonzero sum, noncooperative, two-person game has generated many books, research papers, and has influenced greatly social science thinking. The story, first told by Albert W. Tucker to a group of psychology majors at Stanford University, is based on a strategic game developed by Merrill Flood and Melvin Dresher of the RAND Corporation. It deals with two supposed partners in crime. Tucker's original version of the problem, given at the 1950 Stanford seminar, titled "A Two-Person Dilemma," is stated below. Now known as the prisoner's dilemma, the analysis of the strategic choices and outcomes for each prisoner has contributed important insights in biology, decision analysis, economics, philosophy, political science, sociology, as well as game theory. The book Luce and Raiffa (1957) highlighted the prisoner's dilemma in its discussion of two-person nonzero-sum noncooperative games and appears to be the source for its subsequent popularity and interest. (We have not been able to answer the related dilemma on whether it is the prisoner's dilemma or the prisoners' dilemma.) [*Games and Decisions*, R. D. Luce, H. Raiffa, John Wiley & Sons, New York, 1957; "The prisoner's dilemma," P. D. Straffin, Jr., *The Journal of Undergraduate Mathematics and its Applications*, 1, 1980, 101–103; *Prisoner's Dilemma*, W. Poundstone, Doubleday, New York, 1992]

A Two-Person Dilemma:

The following is how Tucker originally described the prisoner's dilemma, as contained in his mimeographed handout for his Stanford lecture, Straffin (1980).

Two men, charged with a joint violation of law, are held separately by the police. Each is told that (1) if one confesses and the other does not, the former will be given a reward of one unit and the latter will be fined two units, (2) if both confess each will be fined one unit. At the same time each has good reason to believe that (3) if neither confesses, both will go clear.

This situation gives rise to a simple symmetric two-person game (not zero-sum) with the following table of payoffs, in which each ordered pair represents the payoffs to I and II, in that order:

		II	
		confess	not confess
I	confess	$(-1, -1)$	$(1, -2)$
	not confess	$(-2, 1)$	$(0, 0)$

Clearly, for each man the pure strategy "confess" dominates the pure strategy "not confess." Hence, there is a unique equilibrium point given by the two pure strategies "confess." In contrast with this non-cooperative solution one sees that both men would profit if they could form a coalition binding each other to "not confess."

1950 The first OR journal

Under the auspices of the British OR Club, the first scholarly OR journal, the *Operational Research Quarterly*, was published in March 1950. In 1978, its name was changed to the *Journal of the Operational Research Society*.

Initial Issue (ORG):

OPERATIONAL RESEARCH QUARTERLY

VOL. I NO. 1 MARCH 1950

Contents

Note that this first issue contained only the one paper by Blackett!

1950 Nash equilibrium

While a second year student at Princeton, John F. Nash extended von Neumann's minimax theorem for two-person, zero-sum games to prove that every finite n-person, general sum game has at least one equilibrium outcome in mixed strategies. Nash, along with with John C. Harsanyi and Reinhard Selten, received the 1994 Nobel prize in economics for their pioneering analysis of equilibria in the theory of non-cooperative games. As noted in the book by Harold Kuhn and Sylvia Nasar (2002), Nash's approach to the bargaining problem "... has become the standard way of modeling the outcomes of negotiations in a huge theoretical literature spanning many fields, including labor management bargaining and international trade agreements." ["Equilibrium points in n-person games," J. F. Nash, *Proceedings of the National Academy of Sciences,* 36, 1950, 48–49; *A Beautiful Mind*, S. Nasar, Simon & Schuster, New York, 1998; *The Essential John Nash*, H. W. Kuhn, S. Nasar, editors, Princeton University Press, Princeton, 2002]

(©Nobel Foundation)
John F. Nash

(©Nobel Foundation)
John C. Harsanyi

(©Nobel Foundation)
Reinhard Selten

1950 Dynamic programming

Dynamic programming, developed by Richard Bellman, is an optimization technique for multi-stage decision problems based on the principle of optimality: For any optimal policy, whatever the current state and current decision, the remaining decisions must constitute an optimal policy for the state that results from the current decision. Bellman coined both names: dynamic programming and the principle of optimality. Eric Denardo traces the origins of dynamic programming to the sequential decision problems studied by Abraham Wald, Kenneth Arrow, David Blackwell, and Martin Girshick, as well as Bellman's research on functional equations and inventory policies. But, as Denardo notes: "It was Bellman who seized upon the principle of optimality and, with remarkable ingenuity, used it to analyze hundreds of optimization problems . . . ." [*Sequential Analysis*, A. Wald, John Wiley & Sons, New York, 1947; "Optimal inventory policy," K. J. Arrow, D. Blackwell, M. A. Girshick, *Econometrica*, 17, 1949, 214–244; "On the theory of dynamic programming," R. E. Bellman, *Proceedings of the National Academy of Sciences*, 38, 1952, 716–719; *Dynamic Programming*, R. E. Bellman, Princeton University Press, Princeton, 1957 (Dover reprint 2003); *Dynamic Programming: Models and Applications*, E. V. Denardo, Prentice-Hall, Englewood Cliffs, 1982 (Dover reprint 2003); *Eye of the Hurricane*, R. E. Bellman, World Scientific Publishing, Singapore, 1984; "Richard Bellman on the birth of dynamic programming," S. Dreyfus, *Operations Research*, 50, 1, 2002, 48–51]

The calm of dynamic:

In his autobiography (*Eye of the Hurricane*), Bellman recounts how he settled on "dynamic programming" while he was at the RAND Corporation in 1950. Having chosen the term programming to convey the notion of planning and decision making, Bellman recalls: "I wanted to get across that this was dynamic, this was multi-stage, this was time-varying . . . . Let's take a word that has an absolutely precise meaning, namely dynamic, in the classical physical sense. It also has a very interesting property that it is impossible to use the word, dynamic, in a pejorative sense . . . . It was something that not even a Congressman could object to. So I used it as an umbrella for my activities."

(Courtesy the RAND Corporation)

Richard Bellman
1920–1984

1950 OR in agriculture

In 1946, Charles W. Thornwaite, a consulting climatologist, joined Seabrook Farms, New Jersey. Seabrook was the first company to quick freeze its vegetables. It was an integrated farming company: planting, harvesting, processing, quick freezing, storing, and distribution. Noticing that seven thousand acres of peas were maturing at the same time, thus putting a heavy burden on Seabrook's work force and freezing capacity, Thornwaite studied the growth aspects of peas and developed a climatic calendar that showed when to plant and when to harvest. The calendar was then used to develop a planting schedule that enabled mature peas to be harvested at a rate that was in concert with crew scheduling and factory processing capacity. In 1950, all of Seabrooks crops were planted based on a crop's climatic calendar. ["Operations research in agriculture," C. W. Thornthwaite, *Journal of the Operations Research Society of America*, 1, 2, 1953, 33–38; "Operations research in agriculture," C. W. Thornthwaite, pp. 368–380 in *Operations Research for Management*, J. F. McCloskey, F. N. Trefethen, editors, The John Hopkins University Press, Baltimore, 1954]

1950 ***An Introduction to Probability Theory and Its Applications***, Vol. I, William Feller, John Wiley & Sons, New York

This basic reference helped to introduce early OR researchers (and many students) to probabilistic concepts with applications to Markov chains, renewal theory, random walks, and stochastic processes. The long awaited Volume II was published in 1966.

1950 ***Contributions to the Theory of Games***, Vol. I, Harold W. Kuhn, Albert W. Tucker, editors, Annals of Mathematics Studies 24, Princeton University Press, Princeton

By publishing recent and ongoing research in the mathematical theory of games, especially zero-sum two-person games, this volume contributed greatly in making this field an important "new approach to competitive economic behavior." Its companion volumes – II, Annals of Mathematics Studies 28, 1953; III, Annals of Mathematics Studies 39, 1957; and IV, Annals of Mathematics Studies 40, 1959 – helped to bring the then young field of game theory to maturity. Volume IV contains a "reasonably complete" bibliography with 1,009 entries. If bought at the time of their publication, the total cost for all four volumes would have been $18.00.

5

Mathematical, algorithmic and professional developments of operations research from 1951 to 1956

1951 Blending aviation gasolines

How best to run an oil refinery, and, in particular, how to blend aviation gasolines in an optimal manner are the basic problems of oil companies. It was not until the late 1940s and early 1950s when economists and mathematicians joined together to apply the new ideas of linear programming and related mathematical and computational procedures that optimizing methods were successfully developed for the blending problem and applied to the Philadelphia Refinery of the Gulf Oil Company. Today, such methods and their extensions are used to manage and operate the world's oil refineries. ["Blending aviation gasolines – a study in programming interdependent activities," A. Charnes, W. W. Cooper, B. Mellon, pp. 115–145 in *Proceedings: Symposium on Linear Inequalities and Programming*, A. Orden, L. Goldstein, editors, Headquarters, USAF, Washington, April 1, 1952 (also in *Econometrica*, 20, 2, 1952, 135–159); *Linear Programming: The Solution of Refinery Problems*, G. H. Symonds, Esso Standard Oil Company, New York, 1955; "Abraham Charnes and W. W. Cooper (et al.): A brief history of a long collaboration in developing industrial uses of linear programming," W. W. Cooper, *Operations Research*, 50, 1, 2002, 35–41]

The Dynamic Duo:

Both Abraham Charnes and William Cooper were founding members of TIMS, with Cooper serving as its first president (1954) and Charnes as its seventh (1960). They teamed up in 1950 at Carnegie Institute of Technology (now Carnegie Mellon University) to develop mathematical, statistical and econometric methods for use in managing industrial operations. Their over 40 years of joint work has contributed major advances in linear programming and its extensions, including goal programming and data envelopment analysis.

Abraham Charnes
1917–1992

William W. Cooper

1951 First computer-based simplex algorithm

The general simplex algorithm was coded for the National Bureau of Standards SEAC digital computer under auspices of the USAF's Project SCOOP. The first application solved on the SEAC was a U.S. Air Force programming problem dealing with the deployment and support of aircraft. This deployment model can be described as follows: Given the D-Day availability a_0 of a specified type of combat aircraft, and the additional availabilities $a_1, a_2, \ldots, a_n$ in the succeeding n months, determine how to divide these availabilities between combat and training so as to maximize, in some sense, the sortie effort on one or more phases of the war. The system had 48 equations and 71 variables and was solved in 73 simplex iterations in 18 hours, with accuracy to five decimal places. The 18 hours includes the time to store and access data from the SEAC's new and novel magnetic tape system. [*Linear Programming and Extensions*, G. B. Dantzig, Princeton University Press, Princeton, 1963; "The first linear programming shoppe," S. I. Gass, *Operations Research*, 50, 1, 2002, 61–68]

1951 Operations Research Office report: "Utilization of Negro manpower in the Army"

This was a landmark study by the Operations Research Office that "... provided policy-makers in the U. S. Army with objective arguments in favor of integrated units ...." Soon after the study's preliminary report was submitted, the Army initiated its complete integration policy (July 1951). ["Utilization of Negro manpower in the Army," A. H. Hausrath, *Journal of the Operations Research Society of America*, 1, 2, 1954, 17–30]

1951 Nonlinear programming

The general statement of a nonlinear programming problem is as follows: Minimize $f(x)$, subject to $g_i(x) = (\geqslant) 0$ $(i = 1, \ldots, m)$ where all functions are twice continuously differentiable. In their seminal paper, "Nonlinear programming," Harold W. Kuhn and Albert W. Tucker established the name of the field and the mathematical basis for analyzing such problems. The famous Kuhn–Tucker necessary conditions that a solution to a nonlinear inequality system must satisfy stem from this paper. These conditions are now known as the Karush–Kuhn–Tucker conditions in recognition of earlier (1939) unpublished work by William Karush. ["Nonlinear programming," H. W. Kuhn, A. W. Tucker, pp. 481–492 in *Proceedings of the Second Berkeley Symposium on Mathematical Statistics and Probability*, J. Neyman, editor, University of California Press, Berkeley, 1951]

Harold W. Kuhn

Albert W. Tucker
1905–1995

1951 Corporate operations research

An early, if not the first corporation to establish an internal OR group was Courtaulds, Britain's largest producer of viscose yarns. The group, under the direction of A. W. Swan, focused on economic and technical problems such as the optimal use of bobbins and the optimal length of production runs. In the U.S., consultant organizations such as Arthur D. Little started an OR division whose members worked on problems for Sears, Roebuck, Republic Steel, and Simplex Wire & Cable. ["The origins and diffusion of operational research in the UK," M. Kirby, R. Capey, *Journal of the Operational Research Society*, 49, 4, 1998, 307–326; "Operations research," H. Solow, *Fortune*, 4, 1951, 105–107, 146–148]

1951 Optimal dynamic inventory policy

The (S, s) inventory policy is the following: order when the stock on hand falls to s or below, and then order to raise the stock to S. The work by Kenneth Arrow, Theodore Harris, and Jacob Marschak showed how to determine optimal values of (S, s) for a periodic review system with random demand. It was not known, however, that an optimal policy for such an inventory system necessarily followed the (S, s) form. In 1958, Herbert Scarf set out to prove that this was the case, and found that he had to introduce a condition known as K-convexity on the cost functions to obtain the general result. ["Optimal inventory policy,"

K. J. Arrow, T. E. Harris, J. Marschak, *Econometrica*, 19, 1951, 250–272; "The optimality of (S, s) policies in the dynamic inventory problem," H. Scarf, pp. 196–202 in *Mathematical Methods in the Social Sciences*, K. J. Arrow, S. Karlin, P. Suppes, editors, Stanford University Press, Stanford, 1960; "Inventory theory," H. E. Scarf, *Operations Research*, 50, 1, 2002, 186–191]

1951 Imbedded Markov chains in queueing systems

David G. Kendall made an important methodological advance by using the powerful method of imbedded Markov chains to analyze queueing system. For an $M/G/1$ system, Kendall showed that the embedded queue length process at successive departure moments forms a discrete-time Markov chain. ["Some problems in the theory of queues," D. G. Kendall, *Journal of the Royal Statistical Society*, B, 13, 1951, 151–185; "Stochastic processes occurring in the theory of queues and their analysis by the method of imbedded Markov Chains," D. G. Kendall, *Annals of Mathematical Statistics*, B, 13, 1953, 338–354]

1951 First OR university program

The first OR degree programs (M.S. and Ph.D.) were established at The Case Institute of Technology, Cleveland, Ohio. First graduates: 1955 (M.S.) – Lawrence Friedman, Maurice Sasieni; 1957 (Ph.D.) – Eliezer Naddor, Maurice Sasieni. ["West Churchman and Operations Research: Case Institute of Technology, 1951–1957," B. V. Dean, *Interfaces*, 24, 4, 1994, 5–15]

1951 Symposium on "Linear Inequalities and Programming"

Under the joint sponsorship of the U.S. Department of the Air Force (Project SCOOP) and the National Bureau of Standards, a Symposium on Linear Inequalities and Programming was held in Washington, DC, June 14–16, 1951. "Its purpose was to acquaint technical workers in the field of logistics, theory of games, activity analysis approach to quantitative economics (interindustry relations), and military programming with the results of current research on mathematical tools." Many important aspects of linear programming were first presented at the symposium: "A duality theorem based on the simplex method," George B. Dantzig, Alex Orden; "Application of the simplex method to a variety of matrix problems,"Alex Orden; "Blending aviation gasolines – a study in programming interdependent activities," Abraham Charnes, William W. Cooper, Bob Mellon; "The problem of contract awards," Leon Goldstein; "The personnel assignment problem," D. F. Votaw, Jr., Alex Orden. This symposium is considered to be the 1st Mathematical Programming Symposium. [*Proceedings: Symposium on Linear Inequalities and Programming*, Alex Orden, Leon Goldstein, editors, Headquarters, USAF, Washington, April 1, 1952]

1951 *Activity Analysis of Production and Allocation*, Tjalling C. Koopmans, editor, John Wiley & Sons, New York

This book contains the proceedings of the June 20–24, 1949 Cowles Commission for Research in Economics conference held at the University of Chicago (also known as the

0th Mathematical Programming Symposium). It is noted for being the first general publication dealing with linear programming and contains Dantzig's early papers on the linear programming model, the general and transportation simplex methods, linear programming and game theory, and duality theory, plus related papers by Kenneth Arrow, George Brown, Robert Dorfman, David Gale, Murray Geisler, Tjalling Koopmans, Harold Kuhn, Oskar Morgenstern, Paul Samuelson, Herbert Simon, Albert Tucker, and Marshall Wood.

1951 *The Structure of the American Economy, 1919–1939*, 2nd edition, Wassily W. Leontief, Oxford University Press

This book expanded Leontief's earlier (1941) work of the same name that covered the years from 1919–1929. The 1951 volume brought the ideas of input–output analysis to a wider audience of economists, mathematicians and social scientists. The numerical solution of Leontief systems helped to drive early research in computer-based methods for solving systems of linear equations. ["Experiments and large scale computation in economics," O. Morgenstern, pp. 483–549 in *Economic Activity Analysis*, O. Morgenstern, editor, John Wiley & Sons, New York, 1954]

1951 *The Quality Control Handbook*, Joseph M. Juran, editor, McGraw-Hill, New York

This handbook, by one of the founders of the quality movement, became the classic reference for practitioners of quality engineering and reliability. Its fourth edition appeared in 1988.

1951 "Operations Research," Herbert Solow, pp. 105–106, 146, 148 in *Fortune*, April

This is the first article on Operations Research that appeared in the U.S. popular press. It cover's OR's origins, the influence of OR pioneers George Kimball, Horace C. Levinson and Philip Morse, and related early applications. A subsequent (1956) *Fortune* article by Solow described the increasing number of OR applications in business. ["Operations Research is in business," H. Solow, *Fortune*, February, 1956, 128–131, 148, 151–152, 154, 156]

1952 Lindley's equation

Starting from an elementary relation between the waiting times of customers n and $(n + 1)$ in a general $GI/G/1$ queue, Dennis V. Lindley showed that the waiting times have a limiting distribution. He derived an integral equation of the Wiener–Hopf type for this distribution that goes under his name. ["The theory of queues with a single server," D. V. Lindley, *Proceedings of the Cambridge Philosophical Society*, 48, 1952, 277–289]

1952 MIT Committee on Operations Research established

In recognition of the interest in OR by its faculty and students, MIT appointed Philip M. Morse as chairman of the Committee on Operations Research to coordinate education and research in OR. Starting in 1953, the Committee sponsored 15 yearly summer seminars that helped to bring the latest research and applications to the academic and practice communities. In 1955, under Morse's guidance, the Committee was transformed into a cross-campus Operations Research Center (ORC) that supported graduate students and enabled them to work on OR dissertations acceptable to the student's home department. [*In at the Beginnings: A Physicist's Life*, P. M. Morse, MIT Press, Cambridge, 1977; "Philip M. Morse and the Beginnings," John D. C. Little, *Operations Research*, 50, 1, 2002, 146–148]

The first among many:

The first Ph.D. in OR was earned at the ORC by John D. C. Little in 1955 under the supervision of Morse. His thesis title: "Use of Storage Water in a Hydroelectric System." Little was president of ORSA in 1979, president of TIMS in 1984–1985, and first president of INFORMS in 1995.

Little (2002) describes what it was like to be Morse's student. "I remember his office well. He had a totally inadequate blackboard. I recall it as 2.5 feet by 3 feet and you couldn't write more than one and half equations on it. ... Morse's office also contained a couch opposite the blackboard. It was extraordinarily saggy and uncomfortable. Surely nobody overstayed their leave in his office if he had seated them on the couch. ... He was very friendly but business-like and extremely well organized. I have said he was a salesman, but he was not high pressure. Rather he was reasoned and flexible, but behind his demeanor was a very quick mind."

John D. C. Little

1952 Operations Research Society of America (ORSA) founded

The founding meeting of the Operations Research Society of America (ORSA) was held on May 26–27, 1952 in Harriman, New York, at the Arden House, the former estate of the Harriman family operated by Columbia University for scholarly meetings. It was attended by 71 persons who represented a wide range of business, industrial, academic, consultant, military and other governmental organizations. Philip M. Morse was elected president. The first national meeting of ORSA at which technical papers were presented was held on November 17–18, 1952 at the National Bureau of Standards, Washington, DC. It was attended by over 400 members and guests. ["The founding meeting of the society," T. Page, *Journal of the Operations Research Society of America*, 1, 1, 1952, 18–25]

From the ORSA constitution:

"The object of the Society shall be the advancement of the science of operations research, through exchange of information, the establishment and maintenance of professional standards of competence for work known as operations research, the improvement of the methods and techniques of operations research, and the encouragement and development of students of operations research."

1952 First U.S. OR journal

Volume 1, number 1 of *The Journal of the Operations Research Society of America* was published in November 1952. The first editor was Thornton Paige. Its name was changed to *Operations Research* with the February 1956 issue of volume 4, number 1. It is now published as *Operations Research* by the Institute of Operations Research and the Management Sciences (INFORMS).

Initial Issue (J. ORSA)

Number 1 **November, 1952**

1952 Portfolio analysis

The first formulation of a nonlinear programming model that enables an investor to optimally trade-off between expected return and risk in selecting an investment portfolio is due to Harry M. Markowitz. He received the 1990 Nobel prize in economics, joint with Merton H. Miller and William F. Sharpe, for pioneering work in the theory of financial economics. ["Portfolio selection," H. M. Markowitz, *The Journal of Finance*, 7, 1, 1952, 77–91; *Portfolio Selection, Efficient Diversification of Investments*, H. M. Markowitz, John Wiley & Sons, New York, 1959; "Efficient portfolios, sparse matrices, and entities: A retrospective," H. M. Markowitz, *Operations Research*, 50, 2002, 154–160]

(©Nobel Foundation)
Harry M. Markowitz

(©Nobel Foundation)
Merton H. Miller

(©Nobel Foundation)
William F. Sharpe

1952 Parametric programming

Parametric programming considers linear-programming problems in which (1) the coefficients of the objective function or (2) right-hand side values are linear functions of a parameter. Such problems arose from specific applications and were independently investigated by researchers at Project SCOOP and at the RAND Corporation. Straightforward variations of the simplex method applied to these problems produce solutions that are optimal for ranges of the associated parameter. ["Notes on parametric linear programming," A. S. Manne, RAND Report P-468, The RAND Corporation, Santa Monica, 1953; "The parametric objective function, Part I," T. L. Saaty, S. I. Gass, *Operations Research*, 2, 3, 1954, 316–319; "The parametric objective function, Part II: Generalization," S. I. Gass, T. L. Saaty, *Operations Research*, 3, 4, 1955, 316–319; "The computational algorithm for the parametric objective function," S. I. Gass, T. L. Saaty, *Naval Research Logistics Quarterly*, 2, 1, 1955, 39–45; *Linear Programming: Methods and Applications*, S. I. Gass, McGraw-Hill, New York, 1958]

1952 Product form of the inverse

A major advance in developing and maintaining the inverses required by the simplex method was the proposal by Alex Orden to use the product form of the inverse (PFI). The required inverse is expressed as the product of a sequence of matrices, where the matrices in the sequence are elementary elimination matrices. The PFI was used by William Orchard-Hays on the Card Programmed Calculator (CPC) and in the design of his simplex code for the RAND Corporation's IBM 701 computer. The PFI, when combined with the revised simplex (multiplier) method, greatly improved the computational efficiency of the simplex method. ["Application of the simplex method to a variety of matrix problems," A. Orden, pp. 28–50 in *Proceedings: Symposium on Linear Inequalities and Programming*, A. Orden, Leon Goldstein, editors, Headquarters, USAF, Washington, April 1, 1952; "Notes on linear programming: Part V – Alternate algorithm for the revised simplex method using product form for the inverse," G. B. Dantzig, W. Orchard-Hays, TM-1268, The RAND Corporation,

Santa Monica, November 19, 1953; "The RAND code for the simplex method," William Orchard-Hays, RM1269, The RAND Corporation, Santa Monica, 1954; "History of the development of LP solvers," W. Orchard-Hays, *Interfaces*, 20, 4, 1990, 61–73]

Alex Orden

William Orchard-Hays
1918–1989

1952 UNIVAC I installed in The Pentagon to solve U.S. Air Force linear-programming problems

As part of Project SCOOP, the U.S. Air Force installed the second production unit of the UNIVAC I computer in April 1952. It was formally turned over to the Air Force on June 25, 1952. The UNIVAC simplex code was written by the staff of the Air Force's Mathematical Computation Branch under the direction of Emil D. Schell. ["Project SCOOP," E. D. Schell, *Systems for Modern Management*, xvii, 5, 1953, 7, 8, 35]

How large is large?:

The UNIVAC simplex code could solve linear-programming problems of the order (250 × 500). This was considered large-scale at that time. The UNIVAC had 1000 words of "high-speed" memory. The data were stored in long tubes of mercury that had crystals at each end that bounced the data from one end to the other; external data storage was accomplished by means of magnetic tape.

1952 The Society for Industrial and Applied Mathematics (SIAM) founded

The Society for Industrial and Applied Mathematics supports the interactions between mathematics and other scientific and technological communities to: advance the application of mathematics and computational science to engineering, industry, science, and society; promote research that will lead to effective new mathematical and computational methods and techniques for science, engineering, industry, and society; and provide media for the exchange of information and ideas among mathematicians, engineers, and scientists. William E. Bradley, Jr. was SIAM's first president.

1952 *Introduction to the Theory of Games*, J. C. C. McKinsey, McGraw-Hill, New York

This was the first text that presented the concepts of game theory as developed by von Neumann and Morgenstern, and included a discussion of linear programming and its relationship to two-person zero-sum games.

1952 *Operations Research: A Preliminary Annotated Bibliography*, James H. Batchelor, Case Institute of Technology, Cleveland

This was the first such bibliography in operations research. A second edition extended the references through 1957 and was published in 1959 under the title *Operations Research, An Annotated Bibliography*, Saint Louis University Press, Saint Louis. Batchelor's work is noted by his world-wide search for OR papers, books, and reports. Subsequent volumes in 1962, 1963, and 1964 included material through 1961. A total of 9,838 items were cited by all four volumes.

1953 The Institute of Management Sciences (TIMS) founded

The Institute of Management Sciences (TIMS) was founded in 1953 as an international organization for management science professionals and academics. One reason for creating another operations-research oriented organization was the feeling that ORSA, with its historical roots and early emphasis in military applications, would not be adequately responsive to the management world. In 1951–1952, Melvin Savelson initiated discussions and meetings to explore interest in this idea. TIMS was founded on December 1, 1953, at a meeting at Columbia University, organized by Merrill Flood and David Hertz, and involved about 100 attendees. The first president of TIMS was William W. Cooper; Abraham Charnes, Vice President; and Merrill Flood, President Elect. C. West Churchman was chosen as the founding Editor of *Management Science*, first published in October 1954. As the TIMS' constitution required that the immediate past-president serve as Chairman of the TIMS governing council, Andrew Vazsonyi was elected as the first Past President of TIMS, even though he had never served as president! ["Constitution and by-laws of the Institute of Management Sciences," *Management Science*, 1, 1, 1954, 97–102; "The founding of TIMS," W. W. Cooper, Online History Section of INFORMS, 2002; "History in the making," Peter Horner, *ORMS Today*, 29, October 2002, 30–39; "The founding fathers of TIMS," M. E. Salveson, *ORMS Today*, 30, June 2003, 48–53]

From the TIMS constitution:

"The objects of the Institute shall be to identify, extend, and unify scientific knowledge that contributes to the understanding and practice of management."

1953 The Shapley value

The Shapley value is one possible answer to the important question of finding a fair distribution of payoffs in n-person games. Seeking a general answer to this problem, Lloyd Shapley proposed three axioms to capture the idea of a fair distribution and proved that there is a unique imputation that satisfies all three axioms. His treatment of this subject is often cited as a premier exemplar of the use of the axiomatic method. The *Shapley value* can be interpreted as the average marginal contribution of each player when the grand coalition forms, averaged over all $n!$ ways a coalition can be formed, one player at a time. Shapley and Martin Shubik found an immediate application of the concept to voting systems where the Shapley value measured the *a priori* voting power of an individual. ["A value for n-person games," L. S. Shapley, pp. 307–317 in *Contributions to the Theory of Games*, Vol. 2, H. Kuhn, A. W. Tucker, editors, Princeton University Press, Princeton, 1953; "A method for evaluating the distribution of power in a committee system," L. S. Shapley, M. Shubik, *American Political Science Review*, 48, 3, 1954, 787–792; "Game theory at Princeton, 1949–1955: A personal reminiscence," Martin Shubik, pp. 151–163 in *Toward a History of Game Theory*, E. R. Weintraub, editor, Duke University Press, Durham, 1992]

Cutting the cake:

Martin Shubik described the Shapley value as "... one of the most fruitful solution concepts in game theory. It generalizes the concept of marginal value and it, together with the Nash work on bargaining and the Harsanyi value, has done much in the last thirty years to, illuminate the problems of power and fair division ...."

1953 The RAND logistics program

The RAND logistics department was formed in 1953 as part of the Economics Division, which also included the economics analysis and cost analysis departments. On the recommendation of George Dantzig, Murray A. Geisler, who had worked with Dantzig on Project SCOOP, was recruited in 1954 to head RAND's logistical research program.

Early research dealt with the application of economic theory and notions of cost effectiveness to logistics. A highly fruitful application arose in the analysis of flyaway kits used for Strategic Air Command bombers deployed in overseas bases. According to Geisler: "... kits of spare parts had to be prepackaged and flown overseas in the event of an emergency. The problem was what parts to put into these kits so as to maximize their supply performance, given a prespecified weight limit." The RAND analysts used a technique based on marginal analysis to design the kits and were able to show that their kit compositions were superior to those previously packaged by the Air Force. [*A Personal History of Logistics*, M. A. Geisler, Logistics Management Institute, Bethesda, 1986]

From SCOOP to RAND to LMI:

Murray A. Geisler was a branch chief in Project SCOOP, responsible for formulating mathematical models of the U.S. Air Staff's programs. He joined the RAND Corporation in 1954 and served as director of logistics studies and head of the Logistics Department. In 1976, he accepted a position with the Logistics Management Institute (LMI) in Washington, DC. He was president of TIMS in 1961.

Murray A. Geisler
1917–1985

1953 Classification of queueing systems

The widely used notation for classifying queueing systems is due to David G. Kendall. The basic notation uses three major characteristics of a queueing system: the arrival process, the service time distribution, and the number of servers and is written as A/S/c. A fourth and fifth letter are sometimes employed to indicate the maximum number of customers that can be in the queue or in service (K) and the queue discipline (Q). ["Stochastic processes occurring in the theory of queues and their analysis by the method of imbedded Markov Chains," D. G. Kendall, *Annals of Mathematical Statistics*, B, 13, 1953, 338–354; "Queueing theory," D. P. Heyman, pp. 679–686 in *Encyclopedia of Operations Research and Management Science*, 2nd edition, S. I. Gass, C. M. Harris, editors, Kluwer Academic Publishers, Boston, 2001]

David G. Kendall

1953 OR in railroad classification yards

Roger R. Crane
1921–1992

The Operations Research Department at Melpar, Inc., a subsidiary of the Westinghouse Air Brake Company, was established in March 1952. Led by Roger R. Crane, this group initially focused on railroad operations. One of the early studies used queueing analysis to analyze the delay time for freight cars in a railroad classification yard. Using Monte Carlo simulation, the system was modeled as two queues in series, preceding the inspection and classification operations. The study also investigated improvements in the utilization of switching engineers. Roger Crane served as president of TIMS in 1957. ["Analysis of a railroad classification yard," R. R. Crane, F. B. Brown, R. O. Blanchard, *Journal of Operations Research*, 3, 3, 1955, 262–271]

1953 Operational Research Society (UK) founded

On November 10, 1953, the members of the Operations Research Club in England voted to become the Operational Research Society (ORS) with membership open to any person engaged in operational research. The first chairman of the society was O. H. Wansbrough-Jones.

From the ORS constitution:

The objects for which the Society is established are:

(a) the advancement of knowledge, by fostering, promoting and furthering interest in Operational Research, and for such purpose to arrange and organise lectures, classes, discussion and research projects, and to encourage and arrange for contacts between workers in all relevant fields of enquiry;

(b) the advancement of education by providing facilities for and subsidising and encouraging education and training in operational research, and by endowing, organising or supporting scholarships or educational or training schemes in connection therewith, and to conduct examinations or advise on the content of papers for examinations in the subject.

1953 Revised simplex method

A major advance that improved the computational efficiency of the simplex method was the explicit use of the simplex multipliers and the product form of the inverse. ["Notes on linear programming: Part V – Alternate algorithm for the revised simplex method using product form for the inverse," G. B. Dantzig, W. Orchard-Hays, TM-1268, The RAND Corporation, Santa Monica, November 19, 1953]

1953 The Metropolis method

A common problem in statistical physics is to find the energy and configuration of the state of lowest energy for a system composed of many particles. One approach to finding this equilibrium state is to randomly alter the position of each particle and recalculate the resulting energy. If the energy shows a decrease, the new position is accepted. The procedure continues until energy does not change any further. Nicholas Metropolis, Arianna W. Rosenbluth, Marshall N. Rosenbluth, and Augusta H. Teller modified this procedure when the system has a known temperature. The main change involves accepting a move even if it results in an increased energy. If ΔE is the energy change and T the temperature, the move with $\Delta E > 0$ is accepted with probability $e^{-\Delta E/T}$. This procedure is known as the Metropolis method. Years later, it formed a key ingredient of simulated annealing, an optimization search method. ["Equation of state calculations by fast computing machines," N. Metropolis, A. W. Rosenbluth, M. N. Rosenbluth, A. H. Teller, *Journal of Chemical Physics*, 21, 6, 1953, 1087–1092; "The beginning of the Monte Carlo method," N. Metropolis, pp. 125–130 in *From Cardinals to Chaos: Reflections on the Life and Legacy of Stanislaw Ulam*, N. G. Cooper, editor, Cambridge University Press, New York, 1989]

1953 The Allais paradox

The French economist Maurice Allais proposed decision situations that questioned whether the axioms of utility theory apply in practice. In 1952, Allais presented a number of decision examples to prominent theoretical economists with the results showing that their choices implied an inconsistent preference ordering, i.e., the economists did not behave according to the axioms of utility theory. His results are reported in Allais (1953). Discussions of what has since been termed "The Allais paradox" are given in Savage (1954) and Raiffa (1968). ["Le comportement de l'homme rationnel devant le risque: Critique des postulates et axioms de l'école Americaine," M. Allais, *Econometrica*, 21, 1953, 503–546; *The Foundations of Statistics*, L. J. Savage, John Wiley & Sons, New York, 1954; *Decision Analysis*, H. Raiffa, Addision-Wesley, Reading, 1968]

The Father of modern French economics:

Maurice Allais won the 1988 Nobel prize in economics for his pioneering contributions to the theory of markets and efficient utilization of resources.

(©Nobel Foundation)
Maurice Allais

1953 ***The Theory of Inventory Management***, Thomson Whitin, Princeton University Press, Princeton

This book is an early compendium of basic inventory control methods, theory of the firm, and military applications. The second edition (1957) was expanded to include material published after 1953 by Whitin and coauthors that appeared in *Management Science, Journal of the Operations Research Society, and Naval Research Logistics Quarterly*, plus an article by Whitin and H. Wagner on "Dynamic Problems in the Theory of the Firm."

1953 ***Stochastic Processes***, Joseph L. Doob, John Wiley & Sons, New York

This text was one of the first comprehensive measure-theoretic expositions of stochastic processes. The author's approach is clearly stated: "Probability is simply a branch of measure theory ... and no attempt has been made to sugar-coat this fact." The text is historically important for covering martingales in some detail, as well as results obtained earlier by Doob, Paul Lévy and Jean Ville.

1953 ***An Introduction to Linear Programming***, Abraham Charnes, William W. Cooper, A. Henderson, John Wiley & Sons, New York

This book was the first to give an extended discussion of the economic interpretation of linear programming (using the famous nut-mix problem), coupled with the basic mathematical theory and explanation of the simplex method and duality. It also discusses the perturbation of a linear-programming problem that resolves the issue of degeneracy.

The nut-mix problem of Charnes and Cooper (1953):

A manufacturer wishes to determine an optimal program for mixing three grades [A, B, D] of nuts consisting of cashews [C], hazels [H], and peanuts [P] according to the specifications and prices given in table 1. Hazels may be introduced into the mixture in any quantity, provided the specifications are met. The amounts of each nut available each day and their costs are given in table 2. Determine the pounds of each mixture that should be manufactured each day to maximize the gross return (contribution margin).

Table 1

Mixture	Specifications	Selling price: ¢/pound
A	Not less than 50% cashews Not more than 25% peanuts	50
B	Not less than 25% cashews Not more than 50% peanuts	35
D	No specifications	25

Table 2

Inputs	Capacity: pounds/day	Price: ¢/pound
C	100	65
H	60	35
P	100	25
Total	260	

1953 *An Annotated Bibliography on Operations Research*, Vera Riley, Operations Research Office, The Johns Hopkins University, Chevy Chase

This was an early bibliography of the then new field of operations research. It is divided into four sections: History and Methodology, Military Applications, Industrial Applications, Government Planning. Besides annotation, Riley provides biographical material on a number of the authors. In her forward, Riley states: "Operations Research, the bibliographer believes, is an inevitable, logical step in the development of science. IT IS A SCIENTIFIC MOVEMENT." And, "It was the good fortune of operations research that England, under duress of national emergency and motivated by the need of immediate practical results, placed this methodology in a complementary position to executive authority. Here it has remained to provide administrators with a scientific evaluation of alternative courses of action and a quantitative basis for decisions."

1954 Cutting planes for the traveling salesman problem

In their seminal paper, "The solution of a large-scale traveling salesman problem," George B. Dantzig, D. Ray Fulkerson, and Selmer M. Johnson demonstrated the efficacy of cutting planes. Alan J. Hoffman and Philip Wolfe refer to the paper as "... one of the principal events in the history of combinatorial optimization ... important for both what it did and for the future developments it inspired." This paper solved the 49-city traveling salesman problem by starting with a good solution and adding cuts to the assignment formulation. Dantzig's optimistic notion that only a small number of cuts would be required to rule out non-integer solutions was confirmed: only 25 cuts sufficed to prove optimality. This paper established the importance of cutting planes for integer programs. ["The solution of a large-scale traveling salesman problem," G. Dantzig, D. R. Fulkerson, S. M. Johnson, *Operations Research*, 2, 4, 1954, 393–410; "History," A. J. Hoffman, P. Wolfe, Chapter 1 of *The Traveling Salesman Problem*, E. L. Lawler, J. K. Lenstra, A. H. G. Rinnooy Kan, D. B. Shmoys, editors, John Wiley & Sons, New York, 1985]

1954 *Naval Research Logistics Quarterly* sponsored by the Office of Naval Research

This journal was an early and important outlet for theoretical and applied research that impacted logistics, as well as a wide-range of OR topics. Seymour Selig was the first editor. It is now published by Wiley Interscience under the name *Naval Research Logistics*.

1954 *Management Science*, the journal of The Institute of Management Sciences

Volume 1, number 1 of the TIMS sponsored journal, *Management Science*, was published in October 1954. C. West Churchman was the first editor. It is now published by the Institute of Operations Research and the Management Sciences (INFORMS).

Initial Issue MS:

Papers in volume 1, number 1 of *Management Science*:

"Evolution of a 'science of management' in America," H. F. Smiddy, L. Naum;
"Inventory control research: A survey," T. M. Whitin;
"On bus schedules," J. D. Foulkes, W. Prager, W. H. Warner;
"The stepping Stone method of explaining linear programming calculations in transportation problems," A. Charnes, W. W. Cooper;
"The use of mathematics in production and inventory control," A. Vazsonyi;
"Smooth pattens of production," A. J. Hoffman, W. Jacobs;
"A remark on the smoothing problem," H. Antosiewicz, A. J. Hoffman.

1954 Sequencing and scheduling (Johnson's algorithm)

In their book *Theory of Scheduling*, Richard W. Conway, William L. Maxwell, and Louis W. Miller note: "Probably the most frequently cited paper in the field of scheduling is Johnson's solution to the two-machine flow-shop problem. He gives an algorithm for sequencing n jobs, all simultaneously available, in a two-machine flow-shop so as to minimize the maximum flow time. This paper is important, not only for its own content, but also for the influence it has had on subsequent work." ["Optimal two- and three-stage production schedules with setup times included," S. M. Johnson, *Naval Research Logistics Quarterly*, 1, 1, 1954; *Theory of Scheduling*, R. W. Conway, W. L. Maxwell, L. W. Miller, Addison-Wesley, Reading, 1967 (Dover reprint 2003)]

1954 Max-flow min-cut theorem

A network consists of a set of nodes and a set of arcs connecting these nodes, with two distinguished nodes: a source (origin) node and a sink (destination) node. Goods (oil, freight cars, automobiles) can flow from the source node to the sink node across the arcs. Each arc has a capacity above which goods cannot flow across it. Of interest is the maximum amount (flow) of goods that can be sent through the network from the source node to the sink node. Lester R. Ford, Jr. and Delbert Ray Fulkerson showed how to determine the maximum flow by their famous max-flow min-cut theorem. A cut in a network is a set of arcs such that if the cut-set of arcs is removed from the network then goods cannot flow from the source node to the sink node. The capacity of a cut is the sum of the capacities of the arcs in the cut-set. The max-flow min-cut theorem states: For any network the maximal flow value from the source node to the sink node is equal to the minimal cut capacity. ["Maximal flow through a network," L. R. Ford, Jr., D. R. Fulkerson, RAND Research Memorandum 1400, The RAND Corporation, Santa Monica, 19 November 1954 (also in *Canadian Journal of Mathematics*, 8, 3, 1956, 399–404); "On the history of the transportation and maximum flow problems," A. Schrijver, *Mathematical Programming*, B, 91, 3, 2002, 437–445]

Secret min-cut:

Ford and Fulkerson were introduced to the maximal flow through a network problem by Theodore E. Harris of the RAND corporation who, along with retired General F. S. Ross, had formulated a network model of railway traffic flow. The Harris–Ross work was classified secret as it dealt with the finding of a minimal cut of the railway network that shipped goods from the Soviet Union to Eastern Europe. Their work was declassified in 1999 based on a request by Alexander Schrijver (2002) to the Pentagon. Harris and Ross solved their problem by a heuristic "flooding" technique that greedily pushes as much flow as possible through the network. For their 44 node and 105 arc network, Harris and Ross determined a minimal cut with capacity of 163,000 tons.

1954 Dual simplex method

The original, primal simplex method is initiated with a basic feasible solution and then searches a finite sequence of other basic feasible solutions until one is found that also satisfies optimality conditions. In contrast, the dual simplex method starts with an infeasible but optimal basic solution, that is, the basis satisfies the optimality conditions, but its corresponding primal solution has negative components. The process then searches a finite sequence of optimal basic solutions until a feasible one is found. ["The dual method of solving linear programming problems," C. E. Lemke, *Naval Research Logistics Quarterly*, 1, 1, 1954, 36–47]

1954 Branch and bound

The 1954 traveling salesman problem (TSP) study by George B. Dantzig, Lester Ford, and Ray Fulkerson is considered the earliest work to use the branch and bound approach. The first full-fledged use of branch and bound for solving TSPs is due to W. L. Eastman, whose procedure is based on the subtour elimination constraints. The work of Ailsa H. Land and Alison G. Doig, proposed in 1957 and published in 1960, is considered the origin of branch and bound as a general technique for solving integer programs. The term branch and bound is due to John Little et al. in their classic application of the method to the TSP. ["The solution of a large-scale traveling salesman problem," G. Dantzig, D. R. Fulkerson, S. M. Johnson, *Operations Research*, 2, 4, 1954, 393–410; *Linear Programming with Pattern Constraints*, W. L. Eastman, Ph.D. dissertation, Harvard University, 1958; "An automatic method of solving discrete programming problems," A. H. Land, A. G. Doig, *Econometrica*, 28, 1960, 497–520; "An algorithm for the traveling salesman problem," J. Little, K. Murty, D. Sweeney, C. Karel, *Operations Research*, 11, 6, 1963, 972–989; "A tree-search algorithm for mixed integer programming problems," R. J. Dakin, *The Computer Journal*, 8, 1965, 250–255; "History," A. J. Hoffman, P. Wolfe, Chapter 1 of *The Traveling Salesman Problem*, E. L. Lawler, J. K. Lenstra, A. H. G. Rinnooy Kan, D. B. Shmoys, editors, John Wiley & Sons, New York, 1985]

1954 Semi-Markov processes

A semi-Markov process is a process that changes states in accordance with the transition matrix of a discrete-time Markov chain, but takes a random amount of time between the changes. More precisely, whenever the process enters state i, it will visit state j next with probability $p(i, j)$, and, given that the next state is j, the sojourn time in state i has a known distribution F_{ij}. This is a generalization of continuous-time Markov chains where all sojourn times are independent and exponentially distributed with parameters depending on state i alone. Semi-Markov processes are widely applicable, for instance, in studying the $M/G/1$ queue. The pioneering work on the subject was carried out independently by Paul Lévy and W. L. Smith. In a series of papers, R. Pyke provided an extensive treatment and further development of the subject. ["Processus semi-markoviens," P. Lévy, *Proceedings of the International Congress on Mathematics*, 3, 1954, 416–426; "Regenerative stochastic processes," W. L. Smith, *Proceedings of the Royal Society*, A, 232, 1955, 6–31; "Markov renewal processes: Definitions and preliminary properties," R. Pyke, *Annals of Mathematical Statistics*, 32, 1961, 1243–1259]

1954 First award of the Frederick W. Lanchester prize

This prize, established by the Operations Research Society of America (ORSA), is given each year for the best English paper on OR or reporting on an OR study, identified as such. It was first awarded to Leslie C. Edie for his paper "Traffic delays at toll booths," *Operations Research*, 2, 2, 1954, 107–138. From 1954–1960, the prize was jointly sponsored by ORSA and the Johns Hopkins University. The prize is now awarded each year by the Institute of Operations Research and the Management Sciences (INFORMS) for the best English language paper or book in OR. Edie was president of ORSA in 1972. ["Of horseless carriages, flying machines and operations research: A tribute to Frederick William Lanchester (1868–1946)," J. F. McCloskey, *Operations Research*, 4, 2, 1956, 141–147]

Leslie C. Edie

1954 Corporation for Economic and Industrial Research (CEIR)

Founded in 1954, the Washington, DC based CEIR was one of the first companies that provided a wide-range of computer-based operations research consultation services to government and commercial clients. Its Computer Services Division grew into the largest independent commercial computing center and used its IBM 704 and IBM 709 computers to analyze, among other applications, large-scale Leontief interindustry systems and for solving large-scale linear-programming problems. Its president was the economist Herbert W. Robinson, and its staff, over time, included Harold Fassberg, Saul I. Gass, Eli Hellerman, Jack Moshman, and William Orchard-Hays.

1954 The early status of decision making

Ward Edwards' 1954 paper is a state-of-the-art review of decision theory from 1930 to the early 1950s. It was written to bring the mathematical and economic theory of consumer choice to the attention of psychologists. It has proven invaluable as a source document from which one can review and appreciate the work of the post World War II decision science researchers who came from economics, statistics, mathematics, and operations research. The paper's main sections deal with the theory of riskless choices, the application of the theory of riskless choices to welfare economics, the theory of risky choices, the transitivity of choices, and the theory of games and decision functions. 209 references are listed. ["The theory of decision making," Ward Edwards, *Psychological Bulletin*, 51, 4, 1954, 380–417; "The making of decision theory," P. C. Fishburn, pp. 369–388 in *Decision Science and Technology: Reflections on the Contributions of Ward Edwards*, J. Shanteau, B. Mellers, D. Schum, editors, Kluwer Academic Publishers, Boston, 1999]

Ward Edwards

1954 *Operations Research for Management*, Joseph F. McCloskey, Florence N. Trefethen, editors, The Johns Hopkins University Press, Baltimore

This is the first publication that covered: the history of OR and the relationship between management and the operations researcher (authors include C. Goodeve, L. Henderson, E. Johnson); the methods of OR including statistics, information theory, linear programming, queueing theory, suboptimization, symbolic logic, computers, game theory (authors include R. Ackoff, D. Blackwell, W. Cushen, J. Harrison, C. Hitch, P. Morse); and case histories including the famous studies of "Utilization of Negro manpower in the Army" (A. Hausrath) and "Operations Research in Agriculture" (C. Thornthwaite).

1954 *The Compleat Strategyst*, John Williams, McGraw-Hill, New York (Dover reprint 1986)

This book was the first nontechnical exposition of game theory; it emphasized matrix games and their solution. It was quite popular due to its clear exposition and many examples. It was translated into French, Swedish, Russian, Czech, Dutch, Japanese, Polish and Spanish, and although not written as a text, it was adopted by many universities.

Whose life is it, anyway?:

In the second edition, Williams noted that his example of Russian Roulette was renamed by the Russian translator to American roulette!

John D. Williams

1954 *The Foundations of Statistics*, Leonard J. Savage, John Wiley & Sons, New York (Dover reprint 1972)

Called the Bible of Bayesians, this seminal work provided a rigorous axiomatic foundation and philosophical framework for statistical decision making based on a synthesis of von Neumann–Morgenstern utility approach and de Finetti's calculus of subjective probability. ["The foundations of statistics reconsidered," L. J. Savage, pp. 173–188 in *Studies in Subjective Probability*, H. E. Kyburg, Jr., H. E. Smokler, editors, John Wiley & Sons, New York, 1964]

A subjective choice?:

The International Society for Bayesian Analysis and the ASA Section on Bayesian Statistical Science sponsor an annual Leonard J. Savage Award for an outstanding doctoral dissertation in the area of Bayesian Econometrics and Statistics.

Leonard (Jimmie) Savage
1917–1971

1954 *Theory of Games and Statistical Decisions*, David Blackwell, Max A. Girshick, John Wiley & Sons, New York (Dover reprint 1979)

David Blackwell

Intended as a text for first-year graduate students in statistics, this book uses game theory as a framework for the statistical decision theory developed by Abraham Wald. After reviewing the basic theory of games and von Neumann–Morgenstern theory, the book focuses on statistical games and provides a rigorous mathematical treatment of the subject.

1955 Bounded rationality and satisficing

Neo-classical economic theory assumes economic man makes decisions based on perfect and omniscient rationality. That is, individuals, when making rational choices between possible alternatives, maximize expected utility. In contrast, Herbert A. Simon promulgated the principle of bounded rationality: Humans lack both the knowledge and computational skill required to make choices in a manner compatible with economic notions of objective rationality. According to Simon (1987), "Theories of bounded rationality can be generated by relaxing one or more of the assumptions of subjective expected utility theory." This concept, first introduced in two seminal papers (Simon, 1955, 1956), challenged the fundamental tenets of economic decision making. Simon further argued that the goal of maximizing or finding the best choice must be replaced with the goal of satisficing – the selection of an alternative solution that first meets one's stated aspiration levels. For example, an individual on the job market should accept the first job that has a salary of at least $75,000, provides a comprehensive medical plan, and involves overseas assignments. ["A behavioral model of rational choice," H. A. Simon, *Quarterly Journal of Economics*, 69, 1955, 99–118; "Rational choice and the structure of the environment," H. A. Simon, *Psychological Review*, 63, 1956, 129–138; "Bounded rationality," H. A. Simon, pp. 266–268 in *The New Palgrave: A Dictionary of Economics*, Vol. 1, J. Eatwell, M. Milgate, P. Newman, editors, Macmillan Press, New York, 1987; *Models of My Life*, H. A. Simon, Basic Books, New York, 1991; *Economics, Bounded Rationality and the Cognitive Revolution*, H. A. Simon, M. Egidi, R. Marris, R. Vitale, Edward Elgar Publisher, Aldershot, 1992]

The "prophet of bounded rationality":

Herbert A. Simon became interested in the study of decisions when he was 19. This became the constant theme in his entire research life – he referred to it as his monomania. Simon's remarkably broad interests spanned several disciplines. Modern organization theory, computer science, artificial intelligence, and cognitive science can all claim him as a founding father. His full bibliography lists 27 books and nearly 1000 publications. Simon received the 1978 Nobel prize in economics for his pioneering research into the decision-making process within economic organizations. ORSA awarded him the von Neumann theory prize in 1988.

Herbert A. Simon
1916–2001

1955 Computer-based heuristic problem-solving

The collaboration of Herbert A. Simon and Allen Newell gave birth to computer-based heuristic problem-solving, that is, how to program a computer to be a "thinking machine." Simon had met Newell and J. C. (Cliff) Shaw at the System Research Laboratory of the RAND Corporation. By 1954, Newell and Simon were convinced that the way to study problem-solving was to simulate the process with computer programs that could manipulate symbols. The Newell–Simon–Shaw team implemented this approach and created the Logic Theorist (LT), a computer program that used heuristic rules to prove theorems. LT was the first operational artificial intelligence (AI) program. LT produced the first complete proof of a theorem in *Principia Mathematica* on August 9, 1956. They introduced such fundamental AI concepts as list processing languages, heuristic search, production rules, means-end analysis, and verbal protocols. ["Heuristic problem solving: The next advance in operations research," H. A. Simon, A. Newell, *Operations Research*, 6, 1, 1958, 1–10; *Human Problem Solving*, A. Newell, H. A. Simon, Prentice-Hall, Englewood Cliffs,

1972; *Machines Who Think*, P. McCorduck, W. H. Freeman, San Francisco, 1979; *Models of My Life*, H. A. Simon, Basic Books, New York, 1991; *AI: The Tumultuous History of the Search for Artificial Intelligence*, Daniel Crevier, Basic Books, New York, 1993, 258–263]

The clouded crystal ball:

Simon and Newell (1958) predicted that the following events would happen within the next ten years (counting from 1957):
(1) a digital computer would be the world's chess champion;
(2) a digital computer will discover and prove an important new mathematical theorem;
(3) a digital computer will write music that will be accepted by critics as possessing considerable aesthetic value;
(4) theories in psychology will take the form of computer programs, or of qualitative statements about the characteristics of computer programs.

Allen Newell
1927–1992

1955 Stochastic programming

E. Martin L. Beale
1928–1985

The standard linear-programming problem assumes that all data are deterministic. In contrast, stochastic programming, or programming under uncertainty, assumes that data are subject to random variations. Early work in formulating and solving such problems is due to G. B. Dantzig and E. M. L. Beale. ["Linear programming under uncertainty," G. B. Dantzig, *Management Science*, 1, 3–4, 1955, 197–206; "On minimizing a convex function subject to linear equalities," E. M. L. Beale, *Journal Royal Statistical Society*, B, 2, 1955, 173–184]

1955 The kinematical theory of traffic flow

M. J. Lighthill and G. B. Whitham proposed a model of traffic flow that viewed traffic as a special fluid obeying two key principles: (1) flow conservation and (2) a functional relationship between traffic flow and traffic density. From these principles, they derived the propagation of waves in traffic flow and the queueing caused by obstruction of the traffic movement. This seminal theory has led to numerous applications and adaptations. Denos Gazis (2001) cites it as "One of the earliest and most durable contributions to the understanding of traffic flow." ["On kinematic waves: II. A theory of traffic flow on long crowded roads," M. J. Lighthill, G. B. Whitham, *Proceedings of the Royal Society (London)*, A, 229, 1955, 317–345; "Traffic analysis," D. C. Gazis, pp. 843–848 in *Encyclopedia of Operations Research and Management Science*, 2nd edition, S. I. Gass, C. M. Harris, editors, Kluwer Academic Publishers, Boston, 2001]

1955 The capital budgeting problem

The capital budgeting process involves the selection of an optimal portfolio of investments from a set of available independent or interdependent investment projects, given a budget that precludes the selection of all investments. When the objective function and constraints are linear, the problem reduces to a linear or integer programming problem that can be solved readily. The pure capital rationing problem is a special case that arises when the total amount of capital available for investment is limited, the projects are independent, and there is no lending or borrowing. This problem was introduced by James H. Lorie and Leonard J. Savage. ["Three problems in capital rationing," J. H. Lorie, L. J. Savage, *Journal of Business*, 28, 1955, 229–239; "Investment and discount rates under capital rationing – a programming approach," W. J. Baumol, R. E. Quandt, *Economic Journal*, 75, 1965, 317–329; *Mathematical Programming and the Analysis of Capital Budgeting Problems*, H. M. Weingartner, Markham Publishing, Chicago, 1967]

1955 Hungarian method for the assignment and transportation problems

The structure of the defining equations of the assignment and transportation problems is such that both problems can be solved without recourse to the simplex method. The Hungarian method is based on pre-linear programming results in graph theory and matrices by the Hungarian mathematicians, D. König and E. Egerváry, and is due to Harold W. Kuhn. The method was extended to the transportation problem by J. Munkres. ["The Hungarian method for the assignment problem," H. W. Kuhn, *Naval Research Logistics Quarterly*, 1–2, 1955, 83–97; "Algorithms for the assignment and transportation problems," J. Munkres, *Journal of the Society for Industrial and Applied Mathematics*, 5, 1, 1957, 32–38]

1955 The first international congress on telephone traffic

The "First International Congress on the application of the theory of probability in telephone engineering and administration" was held in Copenhagen, at the suggestion of Arne Jensen. The choice of Copenhagen was meant to honor Agner K. Erlang who produced his seminal queueing theory research while working for the Copenhagen Telephone Company. The proceedings of this conference were influential in establishing probability theory as the pre-eminent methodology in analyzing telephone traffic problems. The second International Teletraffic Congress (ITC) was held in The Hague in 1958. [*Introduction to Congestion Theory in Telephone Systems*, R. Syski, Oliver and Boyd, Edinburgh, 1960]

1955 *Linear Programming: The Solution of Refinery Problems*, Gifford H. Symonds, Esso Standard Oil Company, New York

The work of Abraham Charnes, William W. Cooper and Bob Mellon (1952) and Gifford H. Symonds (1953) introduced linear programming to the oil industry. Symonds (president of TIMS in 1956) wrote the first formal account of the use of linear programming in refinery problems. His book covers such problems as blending aviation gasoline, refinery running plan (selection of crude oils to meet product requirements with maximum profit), and the selection of production rates and inventory to meet variable seasonal requirements. Another influential book that dealt with refinery operations was by Alan S. Manne (1956). A 1956 survey of oil industry applications of linear programming was reported by W. W. Garvin, H. W. Crandall, J. B. John, and R. A. Spellman (1957). They note the importance of having high-speed computers and efficient linear programming codes, in particular, the IBM 704 and the LP code written by William Orchard-Hays and Leola Cutler of RAND and Harold Judd of IBM. ["Blending aviation gasolines – a study in programming interdependent activities," A. Charnes, W. W. Cooper, B. Mellon, pp. 115–145 in *Proceedings: Symposium on Linear Inequalities and Programming*, A. Orden, L. Goldstein, editors, Headquarters, USAF, Washington, April 1, 1952 (also in *Econometrica*, 20, 2, 1952, 135–159); "Linear programming for optimum refinery operations," G. H. Symonds, paper presented at the IBM Petroleum Conference, October 26, 1953; *Scheduling of Petroleum Refinery Operations*, A. S. Manne, Harvard University Press, Cambridge, 1956; "Applications of linear programming in the oil industry," W. W. Garvin, H. W. Crandall, J. B. John, R. A. Spellman, *Management Science*, 3, 4, 1957, 407–430]

Gifford H. Symonds

1955 *A Million Random Digits with 100,000 Normal Deviates*, The RAND Corporation, The Free Press, New York

This table of random numbers was widely used in Monte Carlo simulations. The numbers were produced by a rerandomization of a basic table generated by an electronic roulette wheel. This book was a RAND best seller. ["History of RAND's random digits: Summary," W. G. Brown, pp. 31–32 in *Monte Carlo Method*, A. S. Householder, G. E. Forsythe, H. H. Germond, editors, Applied Mathematics Series, Vol. 12, U.S. National Bureau of Standards, Washington, DC, 1951]

Are they random?

00000	10097 32533 76520 13586 34673 54876 80959 09117 39292 74945
00001	37542 04805 64894 74296 24805 24037 20636 10402 00822 91665
00002	08422 68953 19645 09303 23209 02560 15953 34764 35080 33606
00003	99019 02529 09376 70715 38311 31165 88676 74397 04436 27659
00004	12807 99970 80157 36147 64032 36653 98951 16877 12171 76833
00005	66065 74717 34072 76850 36697 36170 65813 39885 11199 29170
00006	31060 10805 45571 82406 35303 42614 86799 07439 23403 09732
00007	85269 77602 02051 65692 68665 74818 73053 85247 18623 88579
00008	63573 32135 05325 47048 90553 57548 28468 28709 83491 25624
00009	73796 45753 03529 64778 35808 34282 60935 20344 35273 88435
00010	98520 17767 14905 68607 22109 40558 60970 93433 50500 73998

(The RAND Corporation, 1955)

1955 ***An Introduction to Stochastic Processes***, Maurice S. Bartlett, Cambridge University Press, Cambridge, U.K.

One of the first texts on stochastic processes, this book was revised in 1966 and 1978. As Peter Whittle notes: "Its flavor was applied in that it considered ... population and epidemic models. It also considered topics equally important for application and theory, e.g., first passage, and the Markov operator formalism .... These developments have very much associated Bartlett with stochastic processes .... It is then something of a surprise to realize, on looking back, that Bartlett's remarkable contributions in this have been almost incidental to his continuing (and undervalued) inferential interests." ["Applied probability in Great Britain," Peter Whittle, *Operations Research*, 50, 1, 2002, 227–239]

1955 ***Studies in the Economics of Transportation***, Martin J. Beckmann, Charles B. McGuire, Christopher B. Winsten, Yale University Press, New Haven

This seminal book can be viewed as the harbinger of the productive interface between operations research modeling and transportation studies, now known as transportation science.

1956 Trim (cutting stock) problem

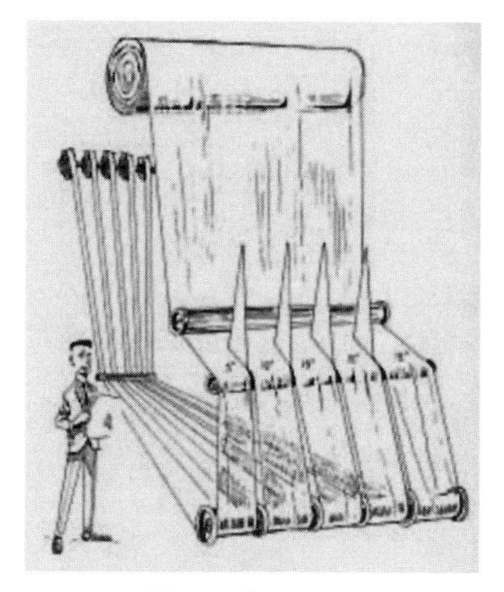

(©Saul I. Gass, 1990)

This is one of the earliest industrial applications of linear programming. It concerns the cutting of standard-width rolls of paper into smaller width rolls to meet the demand for different sizes of cuts while minimizing the trim loss (the left-over rolls whose widths are to small to be used). ["Linear programming: A key to optimum newsprint production," A. E. Paull, *Pulp and Paper Magazine of Canada*, 57, 1, 1956, 85–90; *An Illustrated Guide to Linear Programming*, S. I. Gass, McGraw-Hill, 1970 (Dover reprint 1990)]

1956 Quadratic programming

Many optimization problems (e.g., stock-portfolio selection, structural mechanics, regression analysis, electrical networks) can be formulated mathematically in terms of linear constraints and nonnegative variables, but with an objective function that is quadratic (nonlinear) in the variables. Usually, the objective function is convex and it can then be shown that the problem can be transformed into a linear program and solved by an adaptation of the simplex method. ["An algorithm for quadratic programming," M. Frank, P. Wolfe, *Naval Research Logistics Quarterly*, 3, 1–2, 1956, 95–110; "The simplex method for quadratic programming," P. Wolfe, RAND Report P-1295, The RAND Corporation, Santa Monica, 1957; "Quadratic programming," K. G. Murty, pp. 656–661 in *Encyclopedia of Operations Research and Management Science*, 2nd edition, S. I. Gass, C. M. Harris, editors, Kluwer Academic Publishers, Boston, 2001]

1956 Minimal spanning tree

Given a connected network with *n* nodes and individual costs associated with all edges, the problem is to find a least-cost spanning tree, that is, a subset of edges that connects all nodes and has no cycles, with the sum of its edge costs minimal for all such subsets. This also called the minimal connector/economy tree problem. Efficient algorithms for finding the minimal spanning tree are those by J. B. Kruskal and R. C. Prim. Both are examples of greedy algorithms that lead to optimal solutions. Graham and Pell (1985) give a rather complete history of the problem and discuss earlier algorithmic approaches for solving it. ["On the shortest spanning subtree of a graph and the traveling salesman problem," J. B. Kruskal, *Proceedings of the American Mathematical Society*, 7, 1956, 48–50; "Shortest connection networks and some generalizations," R. C. Prim, *Bell System Technical Journal*, 36, 1957, 1389–1401; *Graphs as Mathematical Models*, G. Chartrand, Prindle, Weber & Schmidt, Boston, 1977; "On the history of the minimum spanning tree problem," R. L. Graham, P. Hell, *Annals of the History of Computing*, 7, 1, 1985, 43–57]

1956 Shortest path problem

Edsger W. Dijkstra published the first efficient algorithm, $O(n^2)$, for the shortest path problem in graphs with n nodes and non-negative edge costs, as well as an algorithm for the shortest spanning tree problem. According to Dijkstra, his shortest path algorithm was "only designed for a demo." The algorithm was intended to demonstrate the power of the ARMAC computer at its official inauguration in Amsterdam in 1956. During the period 1957–1962, a number of shortest path algorithms were proposed. Maurice Pollack and Walter Wiebenson credit the first algorithm to George J. Minty; it had a complexity of $O(n^3)$. Other approaches include those of Richard Bellman, George B. Dantzig, Lester R. Ford, Jr., and E. F. Moore. ["Network flow theory," L. R. Ford, Jr., Paper P-923, The RAND Corporation, July 14, 1956; "A comment on the shortest-route problem," G. J. Minty, *Operations Research*, 5, 5, 1957, 724; "Discrete variable extremum problems," G. B. Dantzig, *Operations Research*, 5, 2, 1957, 266–277; "A variant of the shortest route problem," G. J. Minty, *Operations Research*, 6, 6, 1958, 882–883; "On a routing problem," R. Bellman, *Quarterly Applied Mathematics*, 16, 1958, 87–90; "The shortest path through a maze," E. F. Moore, *Proceedings of an International Symposium on the Theory of Switching*, Part II, April 2–5, 1957, *The Annals of the Computation Laboratory of Harvard University*, Vol. 30, Harvard University Press, Cambridge, 1959; "A note on two problems in connection with graphs," Edsger W. Dijkstra, *Numerische Mathematik*, 1, 1959, 269–271; "Solution of the shortest route problem – a review," M. Pollack, W. Weibenson, *Operations Research*, 8, 2, 1960, 224–230; "An appraisal of some shortest path algorithms," S. E. Dreyfus, *Operations Research*, 3, 1969, 395–412; "EWD1166: From my life," E. W. Dijkstra, pp. 86–92 in *People and Ideas in Theoretical Computer Science*, C. S. Calude, editor, Springer-Verlag, Singapore, 1999]

1956 Pontryagin's maximum principle of optimal control

Lev S. Pontryagin's maximum principle is a necessary condition for the optimal control of a dynamical system governed by the equations $dx/dt = f\{x(t), u(t)\}$ over the time interval $[0, T]$, where $x(t)$ is the state vector with initial value $x(0)$, and $u(t)$ is a control function selected from an admissible set U. The goal is to minimize a cost function $J\{x(T)\}$ that depends on the final value of the state. The maximum principle states that the optimal control $u^*(t)$ maximizes a quantity $H(x^*, p^*, u)$ called the *Hamiltonian* of the system over all controls u in U at every point of the optimal trajectory resulting from the control $u^*(t)$. For linear systems with bounded control variables, the maximum principle implies bang-bang control, implying that the control $u^*(t)$ will flip-flop between extreme values on the boundary of U. The relation between optimal control and nonlinear programming is discussed by Luenberger (1972) and developed by Variaya (1972). ["On the theory of optimal control processes," V. G. Boltyanskii, R. V. Gramkrelidze, L. S. Pontryagin, *Report of the Academy of Sciences of the USSR*, 110, 1, 1956, 7–10; *The Mathematical Theory of Optimal Processes*, L. S. Pontryagin,

Lev S. Pontryagin
1908–1988

V. G. Boltyanskii, R. V. Gramkrelidze, R. V. Mishchenko, John Wiley & Sons, New York, 1962; *Optimal Control*, M. Athans, P. L. Falb, McGraw-Hill, New York, 1965; *Foundations of Optimal Control Theory*, E. B. Lee, L. Markus, John Wiley & Sons, New York, 1967; *Notes on Optimization*, P. P. Varaiya, Van Nostrand Reinhold, New York, 1972; "Mathematical programming and control theory: Trends of interplay," D. G. Luenberger, pp. 102–133 in *Perspectives in Optimization*, A. M. Geoffrion, editor, Addison-Wesley, Reading, 1972]

1956 Société Française de Recherche Opérationelle (SOFRO) founded

The French OR society, SOFRO, was founded in January 1956, with Georges Theodule Guilbaud its first president. In 1964, SOFRO merged with the Association du Droit de l'Informatique et de Traitement de l'Information (AFCALTI) to become the Association Française de l'Informatique et de la Recherche Opérationelle (AFIRO). It is now called the Association Française de Recherche Opérationelle et d'Aide à la Décision (ROADEF).

1956 Arbeitskreis Operational Research (AKOR)

The first German OR Society, the Arbeitskreis Operational Research (AKOR), was founded in 1956 by a group of practitioners, although its membership was open to all. Its first president was Helmut Kregeloh. In 1961, a more theoretical, academic-oriented OR society was formed, the Deutsche Gesellschaft für Unternehmensforschung (DGU) with Henry Görtler as its first president. AKOR and DGU were merged on January 1, 1972 to form the Deutsche Gesellschaft für Operations Research (DGOR) with Hans-Jürgen Zimmerman as its first president. On January 1, 1998, DGOR merged with the Gesellschaft für Mathematik, Ökonometrie und Operations Research (GMÖOR, founded circa 1979) to form the Gesellschaft für Operations Research (GOR) with Peter Kleinschmidt as its first president.

1956 *The Theory of Games and Linear Programming*, Steven Vajda, Methuen & Co., London

This monograph is the first to present a systematic and comprehensive account of the theory of matrix games and linear programming. It was translated into German, French, Japanese and Russian, and helped to introduce these topics in the respective countries and beyond. ["The work of Professor Steven Vajda 1901–1995," K. B. Haley, H. P. Williams, *Journal of the Operational Research Society*, 49, 3, 1998, 298–301]

"Hitler didn't like me very much ... ":

Steven Vajda was born in Budapest, grew up in Vienna, and trained as an actuary before he took his doctorate in mathematics at Vienna University. After Austria capitulated to Hitler in 1938, he and his family managed to emigrate to England. As Vajda said: "Hitler didn't like me very much, but then I didn't like him very much either." During the early months of World War II, he was considered by the British to be enemy alien and was interned for a short while on the Isle of Man where he taught the inmates mathematics. In 1944, he joined the Admiralty Research Laboratory at Teddington, and later became head of Mathematics and Head of OR. Vajda is considered to be the British "father" of linear programming (Haley, Williams, 1998).

Steven Vajda
1901–1995

6

International activities, algorithms, applications, and operations research texts and monographs from 1957 to 1963

1957 First International Conference on operations research

The first international conference in operations research was held at Oxford, England, September 2–6, 1957. It was attended by 250 delegates from 21 countries. It was organized by the OR societies of the U.K., U.S., and Canada. The conference theme was "to unify and extend the science of operational research." The second such international conference, the first sponsored by the International Federation of Operational Research Societies (IFORS), was held in Aix-en-Provence, France, September 5–9, 1960. [*Proceedings of the First International Conference on Operational Research*, M. Davies, R. T. Eddison, T. Page, editors, Operations Research Society of America, Baltimore, 1957; *Proceedings of the Second International Conference on Operational Research*, J. Banbury, J. Maitland, editors, English Universities Press Ltd., London, 1961]

First International Conference on Operations Research – Picture of Attendees

First International Conference on Operations Research – Key to picture of attendees:

First International Conference on Operations Research – List of attendees:
Key to International Conference Photograph

1	A. R. T. Denues	53	Curt Kihlstedt
2	John Harling	54	J. Taylor
3	Macon Fry	55	R. T. Eddison
4	R. P. Hypher	56	Stafford Beer
6	Glen D. Camp	57	Daniel Pigot
7	Charles Salzmann	58	Kenneth Pennycuick
8	Kristen Nygaard	59	A. J. Mayne
9	R. R. Tusenius	60	C. West Churchman
11	Arne Jensen	62	L. J. Mostertman
12	K. B. Haley	63	Johan Philip
13	T. Page	66	C. Berline
14	B. T. Houlden	68	R. A. Leaver
15	R. J. Maher	69	O. Swensson
16	Max Astrachan	70	Edward L. Brink
17	D. R. Read	72	John S. de Cani
18	Miss Alison Doig	73	T. E. Easterfield
19	G. W. Sears	74	Clayton J. Thomas
20	Jan Linderoth	75	E. O. Berdahl
21	L. J. Govier	76	W. J. Reiners
23	W. Monhemius	77	J. K. W. Slater
25	Louis P. Clark	78	Gordon Norton
26	Guy Bitterlin	79	A. H. Schaafsma
27	P. Salmon	80	Gunnar Dannerstedt
30	Jan Sittig	81	Lars Erik Zachrisson
31	LeRoy A. Brothers	82	Jacques A. Zighera
32	Georges Brigham	83	Mrs. A. H. Land
33	Nils Blomqvist	84	E. Kay
34	H. E. Nissen	85	William R. Fair
35	R. W. Bevan	86	Jan Linderoth
36	A. P. MacD. Purdon	87	F. D. Robinson
37	B. D. Hankin	88	Alec Glaskin
38	G. B. Bernard	89	O. Hoflund
39	Georges Parisot	90	Sten Malmqvist
40	T. C. Koopmans	91	Jan Robert Eklind
41	T. Paulsson Frenckner	92	A. Collis-Jones
44	H. K. Weiss	94	Joseph H. Lewis
45	R. A. Acton	95	Erik Holböl Jensen
46	G. M. W. Sebus	96	A. R. v. d. Burg
48	Martin L. Ernst	97	M. P. Barnett
50	E. de Cock	98	Franklin C. Brooks
52	F. W. Santman	99	F. C. Henriques

101	G. E. Nicholson, Jr.	152	R. W. Watkins
102	Max Davies	153	Arne Sjöberg
103	D. G. R. Bonnell	154	S. J. Morrison
104	A. W. Swan	156	J. Stringer
105	Elio M. Ventura	157	Eric Rhenman
106	Robert H. Roy	159	S. P. Rose
107	George J. Feeney	161	Donald Hicks
108	F. J. Toft	163	T. I. McLoughlin
109	M. A. Simpson	165	Uno Fernlund
110	E. D. J. Stewart	166	Alexis G. Joseph
112	Jean Bouzitat	167	Tatsuo Kawata
113	Jean Mothes	168	G. Kreweras
114	Mlle. N. Tabard	169	John Coyle
115	J. G. Wardrop	170	Miss M. A. Frei
117	Hugh J. Miser	171	Ter-Davtian
119	J. J. Wiedmann	172	F. Benson
120	J. Pfanzagl	173	Lyonel Wurmser
121	George Shortley	174	Walter L. Deemer, Jr.
122	Enrique Chacon	175	M. J. Bramson
123	Jan Oderfeld	177	Myhan Erkan
124	Russell L. Ackoff	178	G. N. Gadsby
126	E. H. Palfreyman	179	E. C. Williams
127	R. G. Massey	180	David Valinsky
128	P. G. Smith	182	Jacques Barache
129	J. R. Manning	183	Helmut Kregeloh
130	Andrew Vazsonyi	185	Shiv Kumar Gupta
131	Miss M. S. Munn	186	Anna Restelli
132	M. Solandt	187	Kenichi Koyanagi
135	Anna Cuzzer	188	Miss J. Dinnison
136	G. Nahon	189	A. van Kreveld
137	Alain R. Schlumberger	190	Robert M. Oliver
139	P. Marchand	191	Roger Crane
140	E. de Wilde	192	Bernard O. Koopman
141	J. R. Taylor	193	William Slater
142	Francesco Brambilla	194	Miss B. E. Kornitzer
143	R. S. Gander	195	Charles F. Goodeve
144	W. A. Johnson	196	Thornton Page
145	Edward D. van Rest	197	Philip M. Morse
146	J. R. M. Wanty	198	George B. Dantzig
149	W. E. Duckworth	199	B. H. P. Rivett
150	W. E. Cushen	200	David Bendel Hertz
151	J. L. Venn	201	E. Leonard Arnoff

The *Proceedings* Editors apologize for having failed to recognize all delegates in the photograph.

1957 Operations Research Society of Japan (ORSJ) founded

ORSJ was founded on June 15, 1957. Its first president was Hidesaburo Kurushima.

1957 Project management

Quite often, there are simultaneous, independent scientific investigations of similar problems. The exact date of each development may be somewhat unclear. A case in point is the OR treatment of a problem that came to the forefront during the scientific management investigations of Frederick Taylor and Henry Gantt in the 1900s: how to manage a complex, dynamic project. OR approaches to this problem include: Program Evaluation and Review Technique (PERT); Critical-Path Method (CPM), and the Metra Potential Method (MPM). Each approach has contributed to the real-world management of projects, with variations of these methods being used throughout the world, especially in the construction industry. ["Application of a technique for research and development program evaluation," D. G. Malcolm, J. H. Roseboom, C. E. Clark, W. Fazar, *Operations Research*, 7, 5, 1959, 646–669; "Critical-path planning and scheduling," J. E. Kelley, Jr., M. R. Walker, pp. 160–173 in *Proceedings of the Eastern Joint Computer Conference*, Boston, December 1–3, 1959; "Contribution de la théorie des graphes à l'étude de certains problèmes linéaires," B. Roy, *Comptes rendus des séances de l'Académie des Sciences*, T. 248, séance du 27 avril, 1959, 2437–2439; "Contribution de la théorie des graphes a l'étude des problems d'ordonnancement," B. Roy, pp. 171–185 in *Proceedings of the Second International Conference in Operations Research*, J. Banbury, J. Maitland, editors, English Universities Press Ltd., London, 1961, also see pp. 109–125 in *Les problèmes d'ordonnancement: applications et méthodes*, B. Roy, editor, Dunod, Paris 1964; "A noncomputer approach to the critical path method for the construction industry," J. W. Fondahl, Department of Civil Engineering, Stanford University, Stanford, 1961; "Graphes et ordonnancement," B. Roy, *Revue Française de Recherche Opérationnelle*, 25, 4, 1962, 323–333; *Critical Path Methods in Construction Practice*, J. M. Antill, R. W. Woodhead, Wiley-Interscience, New York, 1965; "Network planning," G. K. Rand, L. Valadares, pp. 561–565 in *Encyclopedia of Operations Research and Management Science*, 2nd edition, S. I. Gass, C. M. Harris, editors, Kluwer Academic Publishers, Boston, 2001]

1957 Quadratic assignment problem

In the standard assignment problem one wants to assign items (people) to locations (jobs) in a manner that optimizes a linear objective function. An important variation arises in facility location where *m* facilities are to be assigned to *m* different locations. For each pair of facilities, a non-negative weight (measure of interaction) is associated with the activity between them. The problem is to place each facility in a location separate from the other facilities in such a way as to minimize the sum of the weights multiplied by the distances between pairs of facilities. ["Assignment problems and the location of economic activities," T. C. Koopmans, M. Beckmann, *Econometrica*, 25, 1957, 53–76; *Facilities Location*, R. F. Love, J. G. Morris, G. O. Wesolowsky, North-Holland, New York, 1988; "Facilities layout," B. K. Kaku, pp. 279–282 in *Encyclopedia of Operations Research and Management Science*, 2nd edition, S. I. Gass, C. M. Harris, editors, Kluwer Academic Publishers, Boston, 2001]

1957 The knapsack problem

The following optimization problem is called the knapsack problem: Maximize $c_1x_1 + c_2x_2 + \cdots + c_nx_n$ subject to $a_1x_1 + a_2x_2 + \cdots + a_nx_n \leqslant b$, with each x_j equal to 0 or 1, with all (a_j, c_j, b) usually taken to be positive numbers. The name is due to interpreting the problem as follows. A camper must select items to be packed in the knapsack from a set of n items, with item j weighing a_j pounds. The camper can carry up to b pounds. But, all items taken together way more than b pounds. The "value" to the camper in selecting item j is given by c_j. The camper wishes to pack a subset of the items so as to maximize the total value. Here, $x_j = 1$ indicates that item j is selected. The knapsack problem, first studied and so named by George B. Dantzig, arises in many industrial and business applications such as selecting a set of projects subject to a budget constraint, and as a subproblem of other problems. ["Discrete-variable extremum problems," G. B. Dantzig, *Operations Research*, 5, 2, 1957, 266–277]

1957 Assignment algorithm is polynomial

James R. Munkres showed that that the Hungarian method for the assignment problem runs in polynomial time, requiring at most $O(n^3)$ operations to solve an $(n \times n)$ assignment problem. Although the significance of this result was not recognized at the time, in the words of Kuhn "it was the first algorithm of polynomial complexity for a large class of linear programs." ["Algorithms for the assignment and transportation problem," J. R. Munkres, *Journal of the SIAM*, 5, 1, 1957, 32–38; "On the origin of the Hungarian method," H. W. Kuhn, pp. 77–81 in *History of Mathematical Programming: A Collection of Personal Reminiscences*, J. K. Lenstra, A. H. G. Rinnooy Kan, A. Schrijver, editors, CWI/North-Holland, Amsterdam, 1991]

1957 Jackson networks

In a queueing network, each node represents a queueing system and flow occurs when customers depart from one node to join the queue at another node. If each node is an $M/M/s$ system and the routing among nodes is Markovian, so that a customer departing node i moves to node j with probability $p(i, j)$, a Jackson network results. Jim Jackson proposed such networks as a means of studying job shops. Despairing of getting useful results from simulation studies, Jackson modeled the job shop as a network of queues and used Markovian assumptions throughout. He then discovered that the long-run solution behaved as if the nodes were treated independently. Starting from this simple model, the analysis of queueing networks grew to be an extremely active and fruitful research area. In the mid 1970s, Frank P. Kelly studied a much larger class of queueing networks that share the two key properties of Jackson networks: the product form of the equilibrium distribution and the Poisson-in-Poisson-out property. ["Networks of waiting lines," J. R. Jackson, *Operations Research*, 5, 4, 1957, 518–521; "Job-shop like queueing systems," J. R. Jackson, *Management Science*, 10, 1, 1963, 131–142; "Networks of queues," F. P. Kelly, *Advances in Applied Probability*, 8, 1976, 416–432; *Reversibility and Stochastic Networks*, F. P. Kelly, John Wiley & Sons, New York, 1979; "How *Networks of Queues* came about," J. Jackson, *Operations Research*, 50, 1, 2002, 112–113]

1957 The perceptron

Frank Rosenblatt, an experimental psychopathologist, proposed a neural net called a perceptron which he simulated on an IBM 704 computer. Rosenblatt's perceptron was simply a layer of McCulloch–Pitts neurons sandwiched between layers of sensor and activation units. Each sensor was a photocell that signaled the amount of light it detected. The McCulloch–Pitts neurons on the second level added up a weighted sum of the sensor signals and fired only if this sum exceeded a threshold. The activation unit translated this fired signal into some form of activity. Rosenblatt's contribution was to adjust the weights to allow the connection between the sensor and neural levels to achieve learning. In their famous 1969 book, *Perceptrons*, Marvin Minsky and Seymour Papert dealt an almost fatal blow to research on neural nets by showing that the perceptron was incapable of carrying out some very basic tasks, such as performing the exclusive-or (XOR) operation. ["The Perceptron: A probablisitc model for information storage and organization in the brain," F. Rosenblatt, *Psychcological Review*, 65, 1958, 386–408; *Principles of Neurodynamics*, F. Rosenblatt, Spartan Books, New York, 1962; *Perceptrons*, M. Minsky, S. Papert, MIT Press, Cambridge, Mass., 1969; *AI: The Tumultuous History of the Search for Artificial Intelligence*, D. Crevier, Basic Books, New York, 1993]

1957 *Dynamic Programming*, Richard Bellman, Princeton University Press, Princeton (Dover reprint 2003)

This book, by the originator of dynamic programming, discusses this important approach to resolving multi-stage decision problems and presents the famous principle of optimality: An optimal policy (set of decisions) has the property that, whatever the initial state and initial decision are, the remaining decisions must constitute an optimal policy with regard to the state resulting from the first decision.

1957 *Games and Decisions: Introduction and Critical Survey*, R. Duncan Luce, Howard Raiffa, John Wiley & Sons, New York (Dover reprint 1989)

This is the first book that "surveys the central ideas and results of game theory and related decision making models without making use of excessive mathematical detail." It made these concepts accessible to practitioners and researchers, especially those working in the behavioral sciences, and helped set the future course of decision making under uncertainty.

1957 *Introduction to Operations Research*, C. West Churchman, Russell L. Ackoff, E. Leonard Arnoff, John Wiley & Sons, New York

This is the first integrated text in operations research written by three OR pioneers who were then associated with the Case Institute of Technology. Although written for the "prospective" consumer and "potential" practitioner, without exercises, it served as a basic text for many years. Churchman and Arnoff were presidents of TIMS (1962 and 1968–1969, respectively), and Ackoff was a president of ORSA (1956).

C. West Churchman
1913–2004

Russell L. Ackoff

E. Leonard Arnoff
1922–1991

1957 **Introduction to Finite Mathematics**, John G. Kemeny, J. Laurie Snell, Gerald L. Thompson, Prentice-Hall, Englewood Cliffs

This text was written for a freshman course in mathematics for students at Dartmouth College. Its aim was to introduce concepts in modern mathematics to students early in their career. It included applications to the biological and social sciences and such new developments as linear programming, game theory and Markov chains, as well as topics in probability theory and vectors and matrices.

1958 Multi-echelon inventory models

Andrew J. Clark coined the term multi-echelon inventory problem to refer to series of inventory locations, each of which receives stock only from the preceding location. Clark initially conducted simulation studies to study multi-echelon systems. He also shared his numerical results on inventory systems with Herbert Scarf. Clark and Scarf's paper on multi-echelon systems used a dynamic programming recursion to solve for the optimal policy. The solution method was based on the key concept of echelon inventory, which assumed fundamental importance in later work on series and assembly systems. ["A dynamic, single-item, multi-echelon inventory problem," A. J. Clark, RM 2297, The RAND Corporation, Santa Monica, 1958; "Optimal policies for a multi-echelon inventory problem," A. J. Clark, H. E. Scarf, *Management Science*, 6, 4, 1960, 475–49; "An informal survey of multi-echelon inventory theory," A. J. Clark, *Naval Research Logistics Quarterly*, 19, 4, 1972, 621–650; "Inventory theory," Herbert E. Scarf, *Operations Research*, 50, 1, 2002, 186–191]

1958 The development of the LISP language

Shortly after its development in 1958 by John McCarthy, the LISP language was universally adopted as the programming language of choice by the community of researchers in artificial intelligence (AI). LISP (which stands for LISt Processing) is a high-level language that is capable of processing lists and replicating the symbol generating and associative capabilities of the human mind. In designing LISP, McCarthy took part of his inspiration from IPL, the language developed by Allen Newell, Herbert Simon, and J. Cliff Shaw for the Logic Theorist at Carnegie-Mellon University. After the 1970s, LISP was widely used in heuristic search techniques and expert systems. [*AI: The Tumultuous History of the Search for Artificial Intelligence*, Daniel Crevier, Basic Books, New York, 1993]

1958 Integer programming and cutting planes

A major theoretical and computational advance for solving integer-programming problems was due to Ralph E. Gomory when he showed how to modify the defining linear problem with a sequence of "cutting planes" (constraints) whose solution, by the simplex method, would thus converge to an integer optimal solution. ["Essentials of an algorithm for integer solutions to linear programs," R. E. Gomory, *Bulletin of the American Mathematical Society*, 64, 5, 1958, 275–278; "On the significance of solving linear programming problems with some integer variables," G. B. Dantzig, *Econometrica*, 28, 1, 1960, 30–44; "An all-integer integer programming algorithm," R. E. Gomory, pp. 193–206 in *Industrial Scheduling*, J. F. Muth, G. R. Thompson, editors, Prentice-Hall, Englewood Cliffs, 1963; "Early integer programming," R. E. Gomory, *Operations Research*, 50, 1, 2002, 78–81]

Fractional cuts and tingling toes:

In 1957, as a consultant to the U.S. Navy, Ralph Gomory was exposed to a linear programming application whose solution returned fractional values of its variables, for example, 1.3 aircraft carriers. After spending a week thinking about the problem, he made the insightful connection that the objective function value of an optimal integer solution to the underlying (maximizing) linear-programming problem would be less than or equal to the integer part of the maximum objective-function value of the linear-programming problem. "No sooner had I made this obvious remark to myself than I felt a sudden tingling in two of my left toes ... (Gomory, 2002)." This insight led Gomory to the method of fractional cuts and the proof of its finiteness.

Ralph E. Gomory

1958 Dantzig–Wolfe decomposition

The constraint structure of many large-scale linear-programming problems is formed by independent subsets that are "tied" together by a small set of additional constraints. For example, the subsets may represent a manufacturing company's production facilities, each independently producing, storing and shipping the company's products. The tie-in constraints would then ensure that the products were being produced so as to meet the total demand for each product within the company's budget, labor, storage and shipping restrictions. The Dantzig–Wolfe decomposition is a modification of the simplex method that enables it to be applied to each subset instead of having to process and solve the complete higher dimension problem. This process was of interest and of value in the early 1950s and 1960s when computers were relatively slow and had little high-speed memory. Today, there are no such concerns as computers and software exist that can solve just about any size problem in an acceptable amount of time. As applied to an economy or industrial complex, the decomposition principle reveals important econometric information related to the optimization of the enterprise. ["Decomposition principle for linear programs," G. B. Dantzig, P. Wolfe, P-1544, The RAND Corporation, Santa Monica, 1958, also in *Operations Research*, 8, 1, 1960, 101–111; *Linear Programming and Extensions*, G. B. Dantzig, Princeton University Press, Princeton, 1963]

1958 *Linear Programming and Associated Techniques: A Comprehensive Bibliography*, Vera Riley, Saul I. Gass, Operations Research Office, The Johns Hopkins University, Chevy Chase, 1958

This annotated bibliography of over 1000 items includes articles, books, monographs, reports, and theses that relate to linear, nonlinear, and dynamic programming, and is rather inclusive.

1958 Canadian Operational Research Society (CORS) founded

The founding meeting of the Canadian Operational Research Society (CORS) was held on April 14, 1958 in Toronto. Osmond M. Solandt was its first president.

1958 *Linear Programming: Methods and Applications*, Saul I. Gass, McGraw-Hill, New York (fifth edition 1984; Dover reprint 2003)

This was the first linear programming book written as a text. It was based on an introductory course in linear programming given at the Graduate School, U.S. Department of Agriculture, Washington, DC. The first and subsequent editions were translated into Russian, Spanish, Polish, Czechoslovakian, Japanese, and Greek, and were the first such texts in the respective countries.

The first of many to come:

Saul I. Gass

1958 *Linear Programming and Economic Analysis*, Robert Dorfman, Paul A. Samuelson, Robert M. Solow, McGraw-Hill, New York (Dover reprint 1987)

This is the first book that emphasized the econometric basis of linear programming and its application to a wide range of related topics. It introduced the power of linear programming and its application to business and industry to the economic profession. Samuelson received the 1970 Nobel prize for the scientific work through which he has developed static and dynamic economic theory and actively contributed to raising the level of analysis in economic science; Solow received the 1987 Nobel prize for his contributions to the theory of economic growth.

Robert Dorfman
1916–2002

(©Nobel Foundation)
Paul A. Samuelson

(©Nobel Foundation)
Robert M. Solow

1958 *Queues, Inventory and Maintenance*, Philip M. Morse, John Wiley & Sons, New York

Written by a prime mover of operations research in the U.S., this expository book brings together for the first time key theoretical and applied aspects of queues. It was the first book published in the ORSA *Publications in Operations Research Series* whose editor was David B. Hertz.

1958 *Scientific Programming in Business and Industry*, Andrew Vazsonyi, John Wiley & Sons, New York

This was one of the first texts on operations research and management science, and is the subjective, personal view of OR/MS by a founding member of TIMS and its first past-president. Written for the "businessman, manager, controller, for the marketing, production, and financial executive, and for the student of business," it helped convey the new ideas of OR/MS with a minimum of mathematics.

A real-life mathematician:

Andrew Vazsonyi authored seven textbooks, plus his autobiography *Which Door has the Cadillac: Adventures of a Real-Life Mathematician*, Writer's Club Press, New York, 2002. In it, he describes his lifelong endeavors in applying OR/MS to business and industry and his adventures as a "real-life mathematician." The book relates how he became TIMS' first past-president without ever being a president, and who were Zepartzatt Gozinto and Endre Weiszfeld.

Andrew (Andy) Vazsonyi
1916–2003

1958 ***Readings in Linear Programming***, Steven Vajda, John Wiley & Sons, New York

This book, written for research and managerial personnel, describes a variety of early applications of linear programming. Each example is stated clearly and is worked out using elementary mathematics.

1958 ***La Théorie des Graphes at ses Applications***, Claude Berge, Dunod, Paris (Dover reprint 2001)

This seminal book presents the theory of graphs in a formal and abstract manner. Dénes König is credited with giving the name graph to structures that can be described by a group of points joined by lines or by arrows (pipelines, sociograms, family trees, communication networks). [*The Theory of Graphs and its Applications*, C. Berge, English edition translated by A. Doig, Methuen & Co., London, 1962]

1958 ***Production Planning and Inventory Control***, John F. Magee, McGraw-Hill, New York

Designed as a practical guide to the subject for managers and engineers, this book suppressed mathematical detail in favor of applications and examples. Accordingly, the book abounds in practical rules and guidelines. Topics covered include lot sizes, safety stock, forecasting, production planning, scheduling, and distribution. Much of the concepts and methods were drawn from work of the Operations Research Group at Arthur D. Little, Inc., a consultant company. John Magee was the first full-time member of the OR Group. The book and its coverage set the tone for later texts in operations management. It remains of value for its clear exposition and insights. Magee was president of ORSA in 1966 and president of TIMS in 1971–1972.

1959 International Federation of Operational Research Societies (IFORS) founded

IFORS is dedicated to the development of operational research as a unified science and its advancement in all nations of the world. The founding members were the Operations Research Society of America (ORSA), the British Operational Research Society (ORS) and the French Société Française de Recherche Opérationelle (SOFRO), with other members being national OR societies. Sir Charles Goodeve was elected the first honorary secretary. IFORS sponsors an international OR conference every three years; the first such conference was held in Aix-en-Provence, France, September 5–9, 1960.

1959 Chance-constrained programming

A mathematical-programming problem in which the parameters of the problem are random variables and for which a solution must satisfy the constraints in a probabilistic sense. ["Chance-constrained programming," A. Charnes, W. W. Cooper, *Management Science*, 1, 1959, 73–79; "Deterministic equivalents for optimizing and satisficing under chance constraints," A. Charnes, W. W. Cooper, *Operations Research*, 11, 1, 1963, 19–39]

1959 Vehicle traffic science

Building on the 1950s research of A. Rueschel and L. A. Pipes, Robert Herman and coworkers developed a car-following theory of traffic flow. This theory postulates that when traffic conditions do not permit drivers to pass one another, each driver follows the vehicle in front in such a way as to avoid coinciding with it in either space or time. In 1959, Ilya Prigogine suggested a model of traffic flow analogous to Ludwig Boltzmann's model for gases in statistical mechanics: A stream of traffic is considered as an ensemble associated with certain statistical properties like desired speed. Subsequently, Herman, Prigogine, and associates further developed this model. Herman was president of ORSA in 1980. Prigogine received the 1977 Nobel prize in chemistry for his contributions to non-equilibrium thermodynamics, particularly the theory of dissipative structures. ["Traffic Dynamics: Analysis of stability in car following," R. Herman, E. W. Montroll, R. B. Potts, R. W. Rothery, *Operations Research*, 7, 1, 1959, 86–106; "Car-following theory of steady-state traffic flow," D. C. Gazis, R. Herman, R. B. Potts, *Operations Research*, 7, 4, 1959, 499–505; "A Boltzmann-like approach to the statistical theory of traffic flow," I. Prigogine, pp. 158–164 in *Proceedings of the First International Symposium on the Theory of Traffic Flow*, R. Herman, editor, 1961; *Kinetic Theory of Vehicular Traffic*, I. Prigogine, R. Herman, American Elsevier, New York, 1971; "Traffic analysis," Denos Gazis, pp. 843–848 in *Encyclopedia of Operations Research and Management Science*, 2nd edition, S. I. Gass, C. M. Harris, editors, Kluwer Academic Publishers, Boston, 2001]

(©Nobel Foundation)
Ilya Prigogine
1917–2003

Robert Herman
1914–1997

The Big Bang:

The physicists George Gamow and Ralph A. Alpher were the first to propose the theory of an expanding universe. Robert Herman, also a physcist, teamed with them later to propose the theory of residual background radiation and suggested that such radiation still exists with a temperature around 5 Kelvin. In 1978, the Nobel prize in physics was awarded for the "discovery of cosmic microwave background radiation" to Arno A. Penzias and Robert Wilson. The current estimate of the background radiation temperature is 2.725 Kelvin.

1959 Random graphs

Given a fixed number of nodes n, if we pick every possible edge with a fixed probability p independently of other edges, a random graph results. The work of Paul Erdös introduced the analysis of random graphs as general structures. Of special interest was the evolution of such graphs when the size of the graph grows as the parameter p increases. In their fundamental work on the growth of random graphs, Paul Erdös and Alfred Rényi derived remarkable results on the distribution of the components of such graphs. As noted by Albert-László Barabási, the 1951 paper of Ray Solomonoff and Anatol Rapoport was a precursor of the Erdös–Rényi work, even though its heuristic derivations were no match for the mathematical elegance of the Erdös–Rényi results. Interest in random graphs has been revived as analysts investigate models of natural growth (such as scale-free networks) to study the underlying architecture of complex networks such as the world-wide web. ["Graph theory and probability," P. Erdös, *Canadian Journal of Mathematics*, 11, 1959,

34–38; "On the evolution of random graphs," P. Erdös, A. Rényi, *Publications of the Mathematical Institute of the Hungarian Academy of Sciences*, 5, 1960, 17–61; "Graph theory and probability – II," *Canadian Journal of Mathematics*, 13, 1961, 346–352; *Linked*, A.-L. Barabási, Penguin Group, New York, 2003; "Scale-free networks," A.-L. Barabási, E. Bonabeau, *Scientific American*, 288, 5, 2003, 60–69]

The traveling mathematician:

Considered to be one of the top 10 mathematicians of the 20th century, the Hungarian Paul Erdös was a child prodigy who, at the age of three, could multiply three-digit numbers in his head and "discovered" negative numbers. He was a citizen of the world who seemed to be in continuous travel as he attended meetings and visited colleagues, often unannounced. He authored and co-authored over 1500 papers.

(©George Csicsery)
Paul Erdös
1913–1996

1959 *Mathematical Methods of Operations Research*, Thomas L. Saaty, McGraw-Hill, New York (Dover reprint 1988)

Thomas L. Saaty

This is the first graduate level text that presented the basic mathematical aspects of OR as applied in optimization, linear programming, game theory, probability, statistics, queueing, with applications and problems. It is noted for its chapter "Résumé of Queueing Theory."

1959 *Mathematical Methods and Theory in Games, Programming and Economics*, Volumes I and II, Samuel Karlin, Addison-Wesley, Reading (Dover Phoenix Series reprint 2003)

The concepts of game theory, mathematical programming and mathematical economics are synthesized in a rigorous unified manner. It was originally published in two volumes, but combined into one volume in the Dover reprint.

1959 *Statistical Forecasting for Inventory Control*, Robert G. Brown, McGraw-Hill, New York

This influential book covered forecasting and stochastic inventory control. It formed the basis of all subsequent decision rules for inventory control and the forecasting methods associated with this subject. The author intended it as a text targeted towards managers, industrial engineers, and OR specialists. In his introduction, Brown offers opposing views on using the past to judge the future by Edmund Burke and Patrick Henry, and goes on to state: "Burke, in modern reincarnation, would very likely be a sales manager, concerned with predictions. Patrick Henry might be a statistician, making forecasts. The astute businessman recognizes the merits of both points of view and realizes that somehow they must be reconciled."

1959 *Testing Statistical Hypotheses*, Erich L. Lehmann, John Wiley & Sons, New York

This important text started as mimeographed class notes (Berkeley, 1949). As a student of Jerzy Neyman, Lehmann was strongly influenced by the Neyman–Pearson treatment of hypothesis testing. He used this theory along with Wald's general decision theory as the theoretical foundations for the text. Originally 400 pages long, the text grew to 600 pages in its second edition (1986). ["Testing statistical hypotheses: The story of a book," E. L. Lehmann, *Statistical Science*, 12, 1, 1997, 48–52]

1959 *The Analysis of Variance*, Henry Scheffé, John Wiley & Sons, New York

This book consolidated and combined analysis of variance theory and methods with Scheffé's related research results. Having spent the first phase of his research mainly on the theoretical aspects of mathematical statistics, Scheffé turned to the analysis of variance and issues of statistical methodology in the 1950s. C. Daniel and Erich Lehmann (1979) note that "the long life of the first edition ... must be attributed to its combination of thoroughness and generality ... with its intuitive and practical insights." ["An analysis of variance for paired comparisons," *Journal of the American Statistical Association*, 47, 1952, 381–400; "Henry Scheffé 1907–1977," C. Daniel, E. L. Lehmann, *Annals of Statistics*, 7, 6, 1979, 1149–1161]

1959 *The Theory of Value*, Gerard Debreu, John Wiley & Sons, New York

(©Nobel Foundation)
Gerard Debreu

This book has the subtitle: "An Axiomatic Analysis of Economic Equlibrium." It presents, for the first time in book form, complete rigorous treatments of the theories of producers' and consumers' behavior, Walrasian equilibrium, Paretian optimum, and their extensions to uncertainty. Debreu was awarded the 1983 Nobel prize in economics for having incorporated new analytical methods into economic theory and for his rigorous reformulation of the theory of general equilibrium.

1960 Planning, Programming and Budgeting System (PPBS)

As first adopted by the U.S. Department of Defense, the concept of a Planning, Programming and Budgeting System (PPBS) deals with an organization setting objectives, the determination of options to achieve these objectives, and the ability to analyze the requisite information, especially cost, that enables the options to be ranked in terms of their effectiveness. PPBS was extended to all government agencies in 1965 by President Lyndon Johnson. By the mid-1970s, the use of PPBS had fallen by the wayside (due to the lack of good data, support and resources), although its ideas on how to improve policy decision making are still quite influential. [*Analysis for Public Decisions*, 3rd edition, E. S. Quade, North-Holland, New York, 1989; "Cost analysis," S. J. Balut, T. R. Gulledge, pp. 152–155 in *Encyclopedia of Operations Research and Management Science*, 2nd edition, S. I. Gass, C. M. Harris, editors, Kluwer Academic Publishers, Boston, 2001]

1960 Airline Group of the International Federation of Operations Research Societies (AGIFORS)

The airline industry formed this special interest group of IFORS in recognition of the value that OR/MS techniques have to the field. The annual symposia of AGIFORS document airline industry applications of OR/MS. [*Airline Operations Research*, D. Teodorovich, Gordon and Breach, New York, 1988; "Airline Operations Research," T. M. Cook, *Interfaces*, 19, 4, 1989, 1–74]

1960 *International Abstracts in Operations Research* (*IAOR*)

Sponsored by the International Federation of Operational Research, IAOR gathers relevant OR articles from some 150 journals and classifies them by processes, applications and methodologies. For each article, IAOR provides an abstract, with title, author, bibliographic information and keywords.

1960 ***Finite Markov Chains***, John G. Kemeny, J. Laurie Snell, D. Van Nostrand Company, Inc., New York

This is the first English language presentation of finite Markov chains. Designed as an undergraduate text, it describes applications to random walks, the Leontief input–output model, and occupational mobility.

1960 ***Stochastic Processes: Problems and Solutions***, Lajos Takács, Methuen, London

Although this concise book is a collection of problems with solutions, it does, however, contain summaries of the main features of the then rather new ideas in Markov chains, Markov processes, and related stochastic processes. Before coming to the United States, Takács' research in Hungary led him to use stochastic models for problems in telephone traffic and the allocation of repairmen in a textile factory. ["Chance or determinism," Lajos Takács, pp. 140–149 in *The Craft of Probabilistic Modeling*, J. Gani, editor, Springer-Verlag, New York, 1986]

1960 ***Dynamic Programming and Markov Processes***, Ronald A. Howard, MIT Press, Cambridge

As described by Ronald A. Howard, the originator of the Markov Decision Process (MDP), "It is based on the Markov process as a system model, and uses an iterative technique similar to dynamic programming as its optimization method." The problem that led to MDP arose around 1957 when Howard was working for the Operations Research Group of the consulting company Arthur D. Little, Inc. It dealt with a decision problem encountered by Sears, Roebuck & Company. The key decision for Sears was to determine whether it was profitable to ship a catalogue to a customer in anticipation of future purchases. Each customer's state was described by the customer's purchase history, where the time period corresponded to a season. Howard developed the MDP method to solve this problem. Although his book has very clear and detailed examples of MDPs, Howard could not then describe the motivating problem due to the proprietary nature of the Sears project. Howard was president of TIMS in 1967. ["Comments on the origin and application of Markov Decision Processes," R. A. Howard, *Operations Research*, 50, 1, 2002, 100–102; "Operations research at Arthur D. Little, Inc.: The early years," J. F. Magee, *Operations Research*, 50, 1, 2002, 149–153]

Ronald A. Howard

1960 ***Introduction to Congestion Theory in Telephone Systems***, Ryszard Syski, Oliver and Boyd, Edinburgh

The purpose of this book was "to present the study of stochastic processes describing the passage of telephone traffic through a switching system, and to introduce to telephone engineers the recent mathematical developments in the general congestion theory applicable to telephone traffic." This authoritative book, the first of its kind, collected all of the previous work on the subject and soon became its bible.

1960 *Planning Production, Inventories, and Work Force*, Charles C. Holt, Franco Modigliani, John F. Muth, Herbert A. Simon, Prentice-Hall, Englewood Cliffs, NJ

One of the highly influential classics in the field, this book defined the area of multi-item production planning. The book grew out of a project funded by the Navy, but the authors frequently referred to a "paint factory" that served as the application site (Pittsburgh Plate Glass Corporation). The centerpiece of the book was the HMMS model that minimized a quadratic cost function, resulting in the celebrated optimal linear decision rules. Another result was the exponentially weighted moving averages model that could handle data with trend and seasonality due to Charles C. Holt, John F. Muth and Peter R. Winters. The book is noted for its clear exposition and for its attention to modeling and validation issues. [*Models of My Life*, Herbert A. Simon, Basic Books, New York, 1991; "Learning how to plan production, inventories, and work force," C. C. Holt, *Operations Research*, 50, 1, 2002, 96–99]

Strange quartet:

In the words of Simon (1991): "The HMMS team harbored simultaneously two Keynesians (Modigliani and Holt), the prophet of bounded rationality (Simon) and the inventor of rational expectations (Muth) – the previous orthodoxy, a heresy, and a new orthodoxy." Simon received the 1978 Nobel prize in economics for his pioneering research into the decision-making process within economic organizations. Modigliani received the 1985 Nobel prize in economics for his pioneering analyses of saving and of financial markets.

(©Nobel Foundation)
Franco Modigliani
1918–2003

1960 *Combinatorial Analysis: Proceedings of the Tenth Symposium in Applied Mathematics*, Richard Bellman, Marshall Hall, Jr., editors, The American Mathematical Society, Providence

This symposium was held at Columbia University, April 24–26, 1958, with its objective to examine the relationship between discrete problems of combinatorial designs and the continuous problems of linear inequalities. Papers discussed (1) the existence and construction of combinatorial designs, (2) combinatorial analysis of discrete extremal problems, (3) problems of communications, transportation and logistics, and (4) numerical analysis of discrete problems. Authors include Richard Bellman, Merrill M. Flood, Murray Gerstenhaber, Ralph E. Gomory, Alan J. Hoffman, Robert Kalaba, Harold Kuhn, Albert W. Tucker.

1961 Simulation programming languages

The first general simulation programming languages were GPSS and SIMSCRIPT. GPSS (General Purpose Simulations System), the first such language, was due to Geoffrey Gordon of IBM; SIMSCRIPT was developed by Harry Markowitz and associates at the RAND Corporation. GPSS uses a highly structured block diagram language and the process interaction form of control. A process is a set of time-oriented events through which an entity must pass. Process interaction control moves an entity through as many events as it can at any given time. In contrast, SIMSCRIPT uses the event scheduling form of control so that a single routine is used to execute all the changes resulting from that event. The SIMSCRIPT mantra, according to Markowitz (2002), states that "the world consists of entities, attributes and sets and changes with events." ["A general purpose systems simulation program," G. Gordon, pp. 87–105 in *Proceedings of the EJCC, Washington, DC*, The Macmillan Company, New York, 1961; "A general purpose systems simulator," G. Gordon, *IBM Systems Journal*, 1, 1, 1962, 18–32; "SIMSCRIPT: A simulation programming language," H. Markowitz, B. Hausner, H. Karr, RM-3310-PR, The RAND Corporation, 1962; "Simulation–computation," G. Gordon, pp. 566–585 in *Handbook of Operations Research: Foundations and Fundamentals*, J. J. Moder, S. E. Elmaghraby, editors, Van Nostrand Reinhold, New York, 1978; *Discrete-Event System Simulation*, J. Banks, J. S. Carson, II, Prentice-Hall, Englewood Cliffs, 1984; *Simulation: A Problem-Solving Approach*, S. V. Hoover, R. F. Perry, Addison-Wesley, Reading, 1989; "Efficient portfolios, sparse matrices, and entities: A retrospective," H. M. Markowitz, *Operations Research*, 50, 1, 2002, 145–160]

1961 Research Analysis Corporation (RAC)

Formed in 1961 as a not-for-profit Federal Contract Research Center, RAC became a key OR analysis organization for the U.S. Army, as well as for many other governmental national security groups. RAC's Army work concentrated on force structure analysis and planning, logistics, military manpower, resource analysis, cost studies, and military gaming. RAC was merged with the General Research Corporation, a for-profit organization, in 1972. ["Operations Research Office and Research Analysis Corporation," E. P. Visco, C. M. Harris, pp. 595–599 in *Encyclopedia of Operations Research and Management Science*, 2nd edition, S. I. Gass, C. M. Harris, editors, Kluwer Academic Publishers, Boston, 2001]

1961 Decision trees

Decision trees were well established in courses taught by Howard Raiffa and Robert O. Schlaifer at the Harvard Business School. Raiffa (2002) recounts: "Because many of our students were bright but mathematically unsophisticated, I formulated most problems in terms of decision trees, which became very standard fare. The standard statistical paradigm was presented in a four-move decision tree ... [involving two pairs of decision and chance nodes] .... I got so used to the use of decision trees in communicating with my students that I couldn't formulate any problem without drawing a decision tree and I was referred to as 'Mr. Decision Tree' ...." The articles by Magee (1964) introduced decision trees to managers and helped to spread their use beyond academia. Decision trees assumed the center stage in Raiffa's 1968 book *Decision Analysis*. ["Decision trees for decision making," J. F. Magee, *Harvard Business Review*, 42, 2, 1964, 126–138; "How to use decision trees for capital investment," J. F. Magee, *Harvard Business Review*, 42, 5, 1964, 79–96; *Decision Analysis*, H. Raiffa, Addison-Wesley, Reading, 1968; "Decision analysis," R. L. Keeney, pp. 423–450 in *Handbooks of Operations Research: Foundations and Fundamentals*, J. J. Moder, S. E. Elmaghraby, editors, Van Nostrand Reinhold, New York, 1978; "Decision analysis: A personal account of how it got started and evolved," H. Raiffa, *Operations Research*, 50, 1, 2002, 179–185]

1961 Little's Law

John D. C. Little provided the first proof of the very useful queueing result: $L = \lambda W$, where L is the average number of customers in the system, W is the average time a customer spends in the system, and λ is the average arrival rate of customers entering the system. This formula is widely applicable since it only requires the queueing system to be ergodic and work preserving. A number of alternative proofs of this result were devised by William Jewell, Shaler Stidham, Jr., and other authors. In Stidham's (2002) view, the great contribution of Little's proof "... was to recast the problem in terms of limiting averages along sample paths, rather than means of limiting or stationary distributions. ... [the proof] mixed sample path arguments with stochastic arguments, the latter exploiting the strict stationarity of the processes involved." This led Stidham to search for a pure sample path proof presented in his 1974 paper which, notwithstanding its title, was not the final word on the subject. ["A proof of the queueing formula: $L = \lambda W$," J. D. C. Little, *Operations Research*, 9, 3, 1961, 383–387; "A simple proof of $L = \lambda W$," W. S. Jewell, *Operations Research*, 15, 6, 1967, 1109–1116; "$L = \lambda W$: A discounted analogue and a new proof," S. Stidham, Jr., *Operations Research*, 20, 6, 1972, 1115–1126, "A last word on $L = \lambda W$," S. Stidham, Jr., *Operations Research*, 22, 2, 1974, 417–421; "Analysis, design, and control of queueing systems," S. Stidham, Jr., *Operations Research*, 50, 1, 2002, 197–216]

1961 Packet switching in computer networks

Packet switching is a procedure for sending data over a computer network. A file of data is subdivided into packets, then numbered, addressed, and forwarded separately to the file's final destination. The packets are stored and forwarded independently of each other through the network based on current capacity and available routings to the destination. When all the packets arrive at the destination, the file is reassembled. The idea was first proposed in 1960–1961 by Paul Baran, an engineer at the RAND Corporation. Baran's interest was based on the need to develop a distributed communications system that would combine survivability with high capacity. In 1962, Leonard Kleinrock of UCLA also suggested that a message be divided it into smaller segments and stored and forwarded. Donald Davies of the National Physical Laboratory, Teddington, England independently arrived at the same idea in 1965. Davies coined the term *packet switching*. In research leading to his doctoral dissertation, Kleinrock developed the theory of stochastic flow of message traffic and the basic principles of packet switching. Packet switching technology formed the basis of the ARPANET, and led to the TCP/IP protocol for the Internet. In his retrospective article, Kleinrock (2002) states: "Packetization helps and is part of today's networking technology, but by itself is not the whole story of the efficiency of networks; rather, the more fundamental gain comes from the introduction of dynamic resource sharing." ["Reliable digital communications systems using unreliable network repeater nodes," P. Baran, Report P-1995, The RAND Corporation, 1960; "Information flow in large communication nets," L. Kleinrock, in *RLE Quarterly Progress Report*, MIT, Cambridge, Mass., 1962; "Message dealy in communication nets with storage," L. Kleinrock, Ph.D. dissertation, MIT, Cambridge, Mass., 1962; *Communication Nets: Stochastic Message Flow and Delay*, L. Kleinrock, McGraw-Hill, New York, 1964 (Dover reprint 1972); "On distributed communications: Introduction to distributed communications networks," P. Baran, RM-3420, The RAND Corporation, 1964; *Computer: A History of the Information Machine*, M. Campbell-Kelly, W. Aspray, Basic Books, New York, 1996, 283–300; *Inventing the Internet*, Janet Abbate, MIT Press, Cambridge, Mass., 1999; "Creating a mathematical theory of computer networks," L. Kleinrock, *Operations Research*, 50, 1, 2002, 125–131]

1961 Goal programming

It is usually assumed that the equations of a linear-programming problem have to be met exactly. For example, in a production–inventory problem, equations usually define how the forecasted weekly demands for a company's products are satisfied in terms of current production and inventory. But, the company may not have enough production capacity to meet all the demands exactly. The demands can then interpreted as achievement levels that may or may not be reached. How to best deviate from these levels can be formulated in terms of a related linear goal-programming problem in which deviations above and below the achievement levels are allowed, but with a penalty. [*Management Models and Industrial Applications of Linear Programming*, Vol. I, A. Charnes, W. W. Cooper, John Wiley & Sons, New York, 1961]

1961 Geometric programming

Many problems in engineering design can be formulated as an optimization problem (geometric program) of the following form: Minimize $g_0(\boldsymbol{t})$ subject to $g_k(\boldsymbol{t}) \leqslant 1$, where $\boldsymbol{t} = (t_1, t_2, \ldots, t_m) > \boldsymbol{0}$ is a vector of variables and, for $k = 0, 1, \ldots, p$, the functions $g_k(\boldsymbol{t})$ are sums of terms having the form $u_i(\boldsymbol{t}) = c_i t_1^{a_{i1}} t_2^{a_{i2}} \cdots t_m^{a_{im}}$ where the sets $\{c_i\} > 0$ and $\{a_{ij}\}$ are arbitrary real numbers. ["A mathematical aid in optimizing engineering design," C. Zener, *Proceedings of the National Academy of Sciences*, 47, 1961, 537–539; *Geometric Programming – Theory and Application*, R. J. Duffin, E. L. Peterson, C. Zener, John Wiley & Sons, New York, 1967; "Geometric programming," Joseph G. Ecker, pp. 330–332 in *Encyclopedia of Operations Research and Management Science*, 2nd edition, S. I. Gass, C. M. Harris, editors, Kluwer Academic Publishers, Boston, 2001]

1961 The theory of auctions

In any auction, the bidders have to navigate between the twin dangers of bidding too high (and paying more than the true value they attach to the item) or too low (so that the item goes to another bidder). In his seminal paper, William Vickrey (1961) introduced an auction (now called a Vickrey auction) where honesty is the best bidding policy. The Vickrey option is a second-price sealed-bid auction where the highest bidder wins the auction but only pays the second highest bid. Vickrey proved that for such an auction, the policy of bidding one's true valuation for the item dominates every other bidding strategy. With the growth of auctions on the Internet, as well as in supply chain and telecommunications bids, the interest in the design and analysis of auctions with desirable economic properties has emerged as an active area of research on the interface of OR and economic theory. Vickrey and James A. Mirrlees won the 1996 Nobel prize in economics for their fundamental contributions to the economic theory of incentives under asymmetric information. ["Counterspeculation, auctions, and competitive sealed tenders," W. Vickrey, *Journal of Finance*, 16, 1961, 8–37; "Auctions and bidding: A Primer," P. Milgrom, *Journal of Economic Perspectives*, 3, 1989, 3–22; *Mathematics and Politics: Strategy, Voting, Power and Proof*, A. D. Taylor, Springer-Verlag, New York, 1995]

1961 *Industrial Dynamics*, Jay W. Forrester, MIT Press, Cambridge

In 1956, Jay W. Forrester gave up his active projects in computer engineering to join MIT's Sloan School of Management as professor of industrial and engineering organization. His new interest in industrial dynamics arose after this transition. The book's theme is that systems of information-feedback control "are fundamental to all life and human behavior." According to Forrester, such systems arise "whenever the environment leads to a decision that results in action which affects the environment and thereby influences future decisions." Forrester's main tool for studying this feedback loop was a deterministic simulation model of a dynamical system governed by the nonlinear system of differential equations $\mathrm{d}\boldsymbol{x}/\mathrm{d}t = \boldsymbol{f}(\boldsymbol{x}, t)$. Systems dynamics, the name that supplanted industrial dynamics, was extended in Forrester's subsequent books to analyze urban growth and decay (*Urban Dynamics*, 1969) and world-wide population and resource issues (*World Dynamics*, 1971). ["Systems dynamics," G. P. Richardson, pp. 807–810 in *Encyclopedia of Operations Research and Management Science*, 2nd edition, S. I. Gass, C. M. Harris, editors, Kluwer Academic Publishers, Boston, 2001]

(Courtesy IEEE)
Jay W. Forrester

1961 *Elements of Queueing Theory*, Thomas L. Saaty, McGraw-Hill, New York (Dover reprint 1983)

This text was one of the first comprehensive treatments of queueing models and applications. It provides an extensive bibliography of 910 references.

1961 *Management Models and Industrial Applications of Linear Programming*, Abraham Charnes, William W. Cooper, Volumes I and II, John Wiley & Sons, New York

Written as a text by two pioneers in operations research and linear programming, these volumes are a collection of the their theoretical research and applied developments. It continues to serve as a source book for OR researchers and graduate students.

1961 *Queues*, D. R. Cox, Walter L. Smith, Methuen & Co., London

This monograph offered the first broad introduction to the theory of queues. It is an introduction to theoretical methods of queues, written from the perspective of an applied mathematician. According to the authors, the intended audience of this monograph was "the operational research worker who is concerned with the practical investigation of queueing." The book starts with Milton's line "They also serve who only stand and wait."

1961 *Applied Statistical Decision Theory*, Robert O. Schlaifer, Howard Raiffa, Division of Research, Harvard Business School, Cambridge (republished, Wiley Classic Library Series, 2000)

Howard Raiffa recalls how, in 1958, the authors of this book set out "to prove that whatever the objectivists could do, we subjectivists could also do – only better...." The key theme of this resource book is to make Bayesianism operational by providing an extensive list of explicit rules for going from prior to posterior distributions. This is accomplished by identifying families of conjugate distributions where the prior and posterior distributions have known forms. A major part of the book is devoted to a list of such distributions. ["Decision analysis: A personal account of how it got started and evolved," H. Raiffa, *Operations Research*, 50, 2002, 179–185]

1962 Benders partitioning method

This is a procedure for solving mixed integer-programming problems of the form maximize $\{cx + dy \mid Ax + By \leqslant b;\ x \geqslant 0 \text{ and integer},\ y \geqslant 0\}$. The problem's structure enables it to be partitioned into related problems with either integer or noninteger variables, but not both. The integer variables are considered to be complicating variables, in that if they were fixed, the resulting problem would be linear. These notions and duality theory leads to a finite convergent algorithm that calls for the repetitive solution of an integer-programming problem and a linear-programming problem. The approach is especially effective if the number of integer variables is small and/or the recurring integer-programming problems are easily solved. Benders' approach has been extended to more general problems in which the variables can be partitioned into two subsets such that the assignment of values to one set reduces the problem to a linear program. ["Partitioning procedures for solving mixed variable programming problems," J. Benders, *Numerische Mathematik*, 4, 1962, 238–252; *Optimization Theory for Large Systems*, L. Lasdon, The Macmillan Company, New York, 1970; *Integer Programming: The-*

ory, Applications, and Computations, H. A. Taha, Academic Press, New York, 1975; *Theory of Linear and Integer Programming*, A. Schrijver, John Wiley & Sons, New York, 1986]

1962 Chinese postman problem

The Chinese mathematician Kwan Mei-Ko first stated the problem of finding a least cost closed traversal of a non-Eulerian graph (a graph that does not contain a cycle that traverses each edge exactly once). Kwan sought to minimize the length of a traversal that included each edge at least once, hence, to find a least cost routing for a postman who must travel (deliver mail) along each edge. This is equivalent to minimizing the sum of the lengths of unproductive moves that correspond to redundant traversal of edges. One can also state this problem as an augmentation problem: what is the least cost way of augmenting a graph (by adding edges) to make it Eulerian? Kwan's paper established necessary conditions of optimality for an augmentation. The key condition was that the cost of added edges along each cycle should not exceed half the cost of the cycle. The precise answer had to wait for the work of Edmonds and Johnson (1973) who applied the minimum-matching algorithm to obtain the augmentation. ["Graphic programming using odd and even points," M.-K. Kwan, *Chinese Mathematics*, 1, 1962, 273–277; "Matching, Euler tours, and the Chinese postman problem," J. Edmonds, E. L. Johnson, *Mathematical Programming*, 5, 1, 1973, 88–124; "Arc routing methods and applications," A. Assad, B. Golden, pp. 375–483 in *Handbooks in Operations Research & Management Science, Vol. 8: Network Routing*, M. O. Ball, T. L. Magnanti, C. L. Momma, G. L. Nemhauser, editors, North-Holland, Amsterdam, 1995; "A historical perspective on arc routing," H. A. Eiselt, G. Laporte, pp. 1–16 in *Arc Routing, Theory, Solutions, and Applications*, M. Dror, editor, Kluwer Academic Publishers, Boston, 2000]

Did the postman ring?:

Kwan is also often referred to as Guan. Alan Goldman suggested the name "Chinese Postman Problem" to Jack Edmonds when Edmonds was in Goldman's Operations Research group at the U.S. National Bureau of Standards. Edmonds appreciated its "catchiness" and adopted it. Goldman was also indirectly influenced by recalling an Ellery Queen detective story called "The Chinese Orange Mystery."

Kwan Mei-Ko

1962 Fuzzy set theory

In the real world, one often encounters sets in which there is no sharp transition from membership in the set to non-membership. For example, the set of "big houses in the neighborhood" is fuzzy in the sense that it does not have crisp boundaries. Given a subset A of X, and a point x in X, the grade of membership of x in A is a function $\mu_A(x)$ that takes on values between 0 and 1. For example, if A denotes the set of big houses and X is the set of all houses in the neighborhood, $\mu_A(x)$ will increase from 0 to 1 as x changes from the smallest house to the largest house in the neighborhood. The membership functions can therefore express attributes of objects in a fuzzy way. The notion of fuzzy sets was introduced by Lotfi A. Zadeh as an alternative to descriptions using probability distributions. Richard Bellman and Zadeh discussed multicriteria decision making using fuzzy sets. A key area of application has been control devices in manufactured goods and fuzzy logic in expert systems. ["From circuit theory to systems theory," L. A. Zadeh, *Proceedings of the Institute of Radio Engineers*, 50, 1962, 856–865; "Fuzzy sets," L. A. Zadeh, *Information and Control*, 8, 1965, 338–353; "Decision-making in a fuzzy environment," R. E. Bellman, L. A. Zadeh, *Management Science*, B, 17, 1970, 141–164; "On the relevance of fuzzy sets in management science methodology," C. Carlson, pp. 11–28 in *Fuzzy Sets and Decision Analysis*, H.-J. Zimmermann, L. A. Zadeh, B. R. Gaines, editors, North-Holland, New York, 1984; *Fuzzy Set Theory*, 2nd edition, H.-J. Zimmermann, Kluwer Academic Publishers, Boston, 1991; *My Life and Travels with the Father of Fuzzy Logic*, Fay Zadeh, TSI Press, 1998]

Lotfi A. Zadeh

1962 Center for Naval Analyses

In 1962, the U.S. Navy reorganized its major contract analysis groups (Institute for Naval Studies and Operations Evaluation Group) into a joint organization, the Center for Naval Analyses, under contract with the Franklin Institute.

1962 *Flows in Networks*, Lester K. Ford, Jr., D. Ray Fulkerson, Princeton University Press, Princeton

This book presented the first unified treatment of the subject. It was most influential in establishing network analysis and related results in graph and combinatorics as OR areas of research and application. It includes detailed discussions of the maximal flow problem and of the "out-of-kilter" method for solving minimal cost network problems.

1962 *Renewal Theory*, D. R. Cox, Methuen & Co., London

This book, the first devoted entirely to renewal theory, provided an accessible and applied treatment of the subject. Peter Whittle describes it as "... virtually a companion volume to the Cox/Smith work on queues..." and notes that "Operations research considerations were foreshadowed in its treatment of failure models and replacement strategies."

The monograph also discusses earlier renewal theory publications by Cox and Smith. ["The superposition of several strictly periodic sequences of events," D. R. Cox, W. L. Smith, *Biometrika*, 40, 1953, 1–11; "On the superposition of renewal processes," D. R. Cox, W. L. Smith, *Biometrika*, 41, 1954, 91–99; "Applied probability in Great Britain," P. Whittle, *Operations Research*, 50, 1, 2002, 227–239]

1962 *The Theory of Probability*, Boris V. Gnedenko, Chelsea Publishing Company, New York

Academician Boris V. Gnedenko was Chair of the Department of Probability Theory, Moscow State University. He belongs to the distinguished line of Russian probabilists and mathematical statisticians being a student of Andrei Kolmogorov (Ph.D. advisor) and Alexander Khintchine. This book is a translation of the second edition of his *Kurs Teorii Veroyatnosteĭ*; it, and its translations, was a long-time standard.

An ambivalent view:

In the Preface to the first Russian edition, Gnedenko states: "The theory of probability is studied as a mathematical discipline exclusively, and the acquisition of specific scientific or engineering results is therefore not at end in itself." Examples and applications are discussed, however, as he notes further: "... probability theory cannot be studied – especially at first – without the systematic solution of problems."

Boris V. Gnedenko
1912–1995

1963 Delphi Method

It is often difficult to have a group or committee of experts converge to a consensus view that they all can agree on. The Delphi Method (process) assumes that the committee is dispersed and communicate their views via a mediator who then coalesces the opinions and arguments into a document that is distributed to all the participants for further discussion. The process may go through several rounds, but it has been found to produce results that can be agreed to by all members of the group. The Delphi Method tends to overcome problems in group dynamics (dominant members, noise, group pressure) by anonymous responses, controlled feedback, and by aggregating individual responses. ["An experimental application of the Delphi Method to the use of experts," N. Dalkey, O. Helmer, *Management Science*, 9, 3, 1963, 458–467; "Delphi method," J. A. Dewar, J. A. Fried, pp. 208–209 in *Encyclopedia of Operations Research and Management Science*, 2nd edition, S. I. Gass, C. M. Harris, editors, Kluwer Academic Publishers, Boston, 2001]

1963 Implicit enumeration

First proposed by Egon Balas, specialized techniques for solving integer-programming problems in which all variables are restricted to be either 0 or 1 have proven to be computationally effective. Here, search algorithms enumerate, either explicitly or implicitly, all 2^n possible solutions. ["Linear programming with zero-one variables" (in Roumanian), E. Balas, *Proceedings of the Third Scientific Session on Statistics*, December 5–7, 1963; "An additive algorithm for solving linear programs with zero-one variables," E. Balas, *Operations Research*, 13, 4, 1965, 517–546; "A multiphase–phase dual algorithm for the zero-one integer programming problem," F. Glover, *Operations Research*, 13, 6, 1965, 879–919; "An improved implicit enumeration approach for integer programming," A. Geoffrion, *Operations Research*, 17, 3, 1969, 437–454; *Integer Programming*, H. M. Salkin, Addison-Wesley, Reading, 1975; *Will to Freedom: A Perilous Journey Through Fascism and Communism*, E. Balas, Syracuse University Press, Syracuse, 2000]

The many lives of Egon Balas:

In his autobiography, Egon Balas (2000) recounts his pre-OR life in Hungary and Romania "As an underground resistance fighter, political prisoner, fugitive, Communist Party official ..." and "... his journey from idealistic young communist to disenchanted dissident." In 1959, at the age of thirty-seven, he decided to transform himself from a Marxist economist to a mathematician. While working on a Romanian forest harvesting programming problem that required some 0–1 variables, he devised the method of implicit enumeration.

Egon Balas

1963 *Smoothing, Forecasting and Prediction*, Robert G. Brown, Prentice-Hall, Englewood Cliffs

Developed by Brown in 1944, exponential smoothing, along with related smoothing and forecasting techniques, are given full discussions in this first such text. A precursor book was Brown's 1959 *Statistical Forecasting for Inventory Control*, McGraw-Hill, New York.

1963 *Analysis of Inventory Systems*, George Hadley, Thomson Whitin, Prentice-Hall, Englewood Cliffs

This text provided a comprehensive treatment of inventory problems with a single stocking point and a single source of supply. Paul Zipkin calls it classic and monumental, the culmination of the first period of research on inventory theory, and notes that the book had a "profound effect on all subsequent developments." For the case of stochastic demand, both reorder point and periodic review systems are covered in detail. The authors start each chapter with a practical and less mathematical discussion of the model. These introductory segments are models of clarity and provide valuable modeling insights. [*Foundations of Inventory Management*, P. H. Zipkin, McGraw-Hill, New York, 2000]

1963 *Computers and Thought*, Edward A. Feigenbaum, Julian Feldman, editors, McGraw-Hill, New York

This collection of 20 articles on artificial intelligence (AI) marked the intellectual debut of the field. According to Feigenbaum, it was meant to provide "a single reference work that summarized the state of the art for the student." The editors purposely selected some of the most readable articles to inform non-specialists about this emergent field. [*Machines Who Think*, P. McCorduck, W. H. Freeman, San Francisco, 1979]

1963 *Linear Programming and Extensions*, George B. Dantzig, Princeton University Press, Princeton

This book, by the "father" of linear programming and the inventor of the simplex method, has served generations of OR analysts and students as a source and text for both theory and applications. It includes most of Dantzig's theoretical and applied developments in linear programming and its extensions up to that time. Dantzig was president of TIMS in 1966.

LINEAR PROGRAMMING AND EXTENSIONS

by GEORGE B. DANTZIG
THE RAND CORPORATION
and
UNIVERSITY OF CALIFORNIA, BERKELEY

1963
PRINCETON UNIVERSITY PRESS
PRINCETON, NEW JERSEY

George B. Dantzig

7

Methods, applications and publications from 1964 to 1978

1964 Complementarity problems

The general complementarity problem is concerned with finding a vector $x = (x_i) \geqslant 0$ that satisfies the inequality system $f_i(x) \geqslant 0$ with $x_i f_i(x) = 0$ $(i = 1, \ldots, n)$. An equivalent formulation is finding a solution (x, y) to the system $y - f(x) = 0$, $x \geqslant 0$, $y \geqslant 0$, $x^{\mathrm{T}} y = 0$. The pair of vectors (x, y) forms a complementary solution if $x_i y_i = 0$ for each pair of components (x_i, y_i) for $i = 1, \ldots, n$. If $f(x) = q + Mx$ the problem is called a linear complementarity problem; otherwise, it is termed nonlinear. The symmetric primal and dual linear-programming problems, the quadratic programming problem, and the bimatrix two-person non-zero-sum game can all be transformed into a linear complementarity problem. Under varying assumptions on the form of the matrix M (e.g., positive semi-definite), algorithms exist for solving linear complementarity problems, especially for bimatrix and quadratic-programming problems. Linear and nonlinear complementarity problems have found application in economics, engineering, game theory, and finance. ["Note on a fundamental theorem in quadratic programming," R. W. Cottle, *SIAM Journal of Applied Mathematics*, 12, 3, 1964, 663–665; "Nonlinear programs with positively bounded Jacobians," R. W. Cottle, Technical Report ORC 64-12 (RR), Operations Research Center, University of California, Berkeley, 1964, also in *SIAM Journal of Applied Mathematics*, 14, 1, 1966, 147–158; "Equilibrium points of bimatrix games," C. E. Lemke, J. T. Howson, *SIAM Journal of Applied Mathematics*, 12, 2, 1964, 412–423; "Bimatrix equilibrium points and mathematical programming," C. E. Lemke, *Management Science*, 11, 7, 1965, 681–689; "Complementary pivot theory of mathematical programming," R. W. Cottle, G. B. Dantzig, *Linear Algebra and its Applications*, 1, 1968, 103–125; *Linear Complementarity, Linear and Nonlinear Programming*, K. G. Murty, Heldermann-Verlag, Berlin, 1988; "Engineering and economic applications of complementarity problems," M. C. Ferris, J. S. Pang, *SIAM Review*, 39, 4, 1997, 669–713; "Complementarity problems," R. W. Cottle, pp. 115–118 in *Encyclopedia of Operations Research and Management Science*, 2nd edition, S. I. Gass, C. M. Harris, editors, Kluwer Academic Publishers, Boston, 2001]

1964 Vehicle routing savings algorithm

The basic vehicle routing problem deals with the assignment of centrally located vehicles to delivery routes so that the total cost of the assignment is minimized. The cost is assumed to be a direct function of the miles driven; the trucks can have different load capacities. An approximate, but, in general, a near-optimal solution to the problem is given by the savings algorithm of G. Clarke and J. W. Wright. ["Scheduling of vehicles from a central depot to a number of delivery points," C. Clarke, J. W. Wright, *Operations Research*, 12, 4, 1964, 568–581]

1964 *Analysis for Military Decisions*, Edward S. Quade, editor, North-Holland, Amsterdam

This was the first of several RAND books authored by Edward S. Quade, a mathematician who played a major role in developing the framework and methodology of systems analysis for use in military and civilian decision making. This book contains and extends material of a course "An Appreciation of Analysis for Military Decisions" given at the RAND Corporation in 1955 and 1959. A key focus of the book is systems analysis, defined by Quade as "OR applied to the determination of force posture – that is, to the selection of future weapons systems and to the management of the process for developing and acquiring these weapons." It includes, among others, chapters by Charles J. Hitch (Analysis for Air Force decisions), Quade (Mathematics and systems analysis) and Thomas C. Schelling (Assumptions about enemy behavior). ["Military systems," E. S. Quade, pp. 503–534 in *Handbook of Operations Research: Models and Applications*," J. J. Moder, S. E. Elmaghraby, editors, 1978; "RAND Corporation," G. H. Fisher, W. E. Walker, pp. 690–695 in *Encyclopedia of Operations Research and Management Science*, 2nd edition, S. I. Gass, C. M. Harris, editors, Kluwer Academic Publishers, Boston, 2001]

1964 *Studies in Subjective Probability*, Henry E. Kyburg, Jr., Howard E. Smokler, editors, John Wiley & Sons, New York

This collection of articles represents some of the most important historical research in the field of subjective probability. The authors include John Venn (1888), Émile Borel (1924), Frank P. Ramsey (1926), Bruno de Finetti (1937), Bernard O. Koopman (1940), and Leonard J. Savage (1961). The editors provide a useful introduction. The article by de Finetti is a translation of *La prévision* (*Foresight: Its Logical Laws, Its Subjective Sources*) with new notes by the author.

1964 *Decision and Value Theory*, Peter C. Fishburn, John Wiley & Sons, New York

The subject of this book is a prescriptive theory of choice for individual decisions. Combining and extending earlier approaches to estimating the relative values of objectives with subjective probability, this book provides a seminal view of procedures that deal with decision making in terms of expected relative values or expected utilities of strategies.

1964 *Monte Carlo Methods*, John M. Hammersley, David C. Handscomb, Methuen & Co., London

This book was the first comprehensive text dealing with the development and use of Monte Carlo methods to solve simulation problems.

1964 ***Principles of Random Walk***, Frank L. Spitzer, Van Nostrand, Princeton

This book was the first comprehensive graduate-level treatment of random walks. Spitzer used Wiener–Hopf techniques to reveal connections between the analysis of the single-server queue and random walks. Certain key measures of interest, such as the waiting time of an arriving customer, can be related to maximum or minimum functionals of the random walk associated with the queue. Subsequent research showed that the combinatorial techniques used to study random walks were equally fruitful when applied to queueing systems.

1965 Surrogate constraints

For a 0–1 integer-programming problem with a set of $(m \times n)$ constraints $Ax \leqslant b$, a surrogate constraint is defined as follows: Let u be an m-dimensional nonnegative multiplier vector and form the surrogate constraint $u(A)x \leqslant ub$. Note that not all components of u need be positive, thus, the surrogate constraint is a non-negative combination of some or all of the given constraints. Fred W. Glover showed that such constraints, when added systematically to the original problem, cause enumerative and branch-and-bound solution procedures to be more efficient in their search for an optimal integer solution. The application of such constraints can be carried over to general integer-programming problems. ["A multiphase-dual algorithm for the zero-one integer programming problem," F. Glover, *Operations Research*, 13, 6, 1965, 879–919; "Surrogate constraints," F. Glover, *Operations Research*, 16, 4, 1968, 741–749]

1965 Complexity theory

Jack Edmonds

In his path-breaking paper on the matching problem, Jack Edmonds raised the broader issue of what is a "good" algorithm. Such algorithms are efficient in the sense of having worst-case computational complexity that increases only as a polynomial in the size of the problem. He wrote: "I am claiming, as a mathematical result, the existence of a good algorithm for finding a maximum cardinality matching in a graph. There is an obvious finite algorithm, but that algorithm increases in difficulty exponentially in the size of the graph. It is by no means obvious whether or not there exists an algorithm whose difficulty increases only algebraically with the size of the graph." He went on to comment that he was not prepared "to set up the formal machinery" to give these statements formal meaning. In the early 1970s, the work of Stephen A. Cook and Richard M. Karp formalized the good algorithm concept and initiated the theory of algorithmic complexity analysis. ["Paths, trees, and flowers," Jack Edmonds, *Canadian Journal of Mathematics*, 17, 1965, 449–467; "The complexity of theorem proving procedures," S. A. Cook, *Proceedings of the 3rd ACM Symposium on the Theory of Computing*, ACM, New York, 1971, 151–158; "Reducibility among combinatorial problems," R. M. Karp, pp. 85–103 in *Complexity of Computer Computations*, R. E. Miller, J. W. Thatcher, editors, Plenum Press, New York, 1972; "A glimpse of heaven," J. Edmonds, pp. 32–54 in *History of Mathematical Programming, A Collection of Personal Reminiscences*, J. K. Lenstra, A. H. G. Rinnooy Kan, A. Schrijver, editors, North-Holland, Amsterdam, 1991]

1965 Political redistricting

Based on the decennial census, the 50 U.S. states must define congressional districts that meet constitutional criteria: approximately equal population districts to reflect one-man–one-vote, and contiguous and compact districts. The first OR approach to resolving this problem was done by Sidney W. Hess, J. B. Weaver, H. J. Siegfeldt, J. N. Whelan, and P. A. Zitlau. Similar applications arise in determining salesman territories, police districts, medical response areas, warehouse location. Hess was president of TIMS in 1976–1977. ["Nonpartisan political redistricting by computer," S. W. Hess, J. B. Weaver, H. J. Siegfeldt, J. N. Whelan, P. A. Zitlau, *Operations Research*, 13, 6, 1965, 998–1006; "Optimal political districting by implicit enumeration techniques," R. S. Garfinkel, G. L. Nemhauser, *Management Science*, 16, 8, 1970, B495–B508]

1965 Expert systems

Expert systems use a body of stored knowledge and an inference engine to offer advice on difficult problems. Early expert systems typically had narrow application domains. DENDRAL was devised to assess the global structure of complex organic molecules based on mass spectrometric and other chemical data. The problem was posed in 1965 to Edward Feigenbaum by the Nobel laureate Joshua Lederberg. DENDRAL made use of a graph theoretic algorithm Lederberg had devised in 1964 for generating all possible molecular structures for a compound with certain known chemical properties. Feigenbaum and Robert K. Lindsay headed the DENDRAL project for the next ten years. The key to DENDRAL was the use of expert rules that could drastically reduce the search space so that the remaining candidates could be examined exhaustively, if need be (e.g., from 11 million to about 22,000 possibilities for a specific compound). In the early 1970s, Feigenbaum and Bruce Buchanan started to explore the use of rule-based programs consisting of if–then rules, called production systems. Buchanan and his doctoral student Edward Shortliffe set out to embody practical medical knowledge into a production system. The result was MYCIN, an expert system that suggested the cause of a blood infection and the appropriate antibiotics to be used in its treatment. MYCIN was the first expert system to separate the rules (the knowledge base) from the logic required to apply them (the inference engine). This led to the development of expert system shells. ["On generality and problem solving; A case study using the DENDRAL program," E. Feigenbaum, B. Buchanan, J. Lederberg, pp. 165–190 in *Machine Intelligence*, D. Michie, editor, Elsevier, New York, 1971; *MYCIN: Computer-Based Medical Consultations*, E. H. Shortliffe, Elsevier, New York, 1976; *The Handbook of Artificial Intelligence*, Vol. 2, A. Barr, E. A. Feigenbaum, editors, Addison-Wesley, Reading, 1982; *Rule-Based Expert Systems: The MYCIN Experiments of the Stanford Heuristic*

Programming Project, B. G. Buchanan, E. H. Shortliffe, Addison-Wesley, Reading, 1984; *AI: The Tumultuous History of the Search for Artificial Intelligence*, Daniel Crevier, Basic Books, New York, 1993]

1965 *Queues and Inventories: A Study of Their Basic Stochastic Processes*, Narahari Umanath Prabhu, John Wiley & Sons, New York

This book was one of the first to bring together, under a common framework, a set of seemingly disparate problems of applied probability.

1966 Decision analysis

One of the earliest decision analysis of a practical problem was C. Jack Grayson's 1962 dissertation (under Howard Raiffa at Harvard) of how oil wildcatters made decisions. In 1964, Raiffa taught a graduate course in decision analysis in the economics department and started to prepare materials for a book under that title. At Stanford, Ronald Howard had adopted the name decision analysis for his research program. His 1966 paper is the first published paper referring to decision analysis and its domain of application. The paper by Peter C. Fishburn gives a detailed discussion of how decision theory evolved and contains a rather complete listing of related references. [*Decisions under Uncertainty: Drilling Decisions by Oil and Gas Operators*, C. J. Grayson, Division of Research, Harvard Business School, Cambridge, Mass., 1962; "Decision analysis: Applied decision theory," R. A. Howard, pp. 55–71 in *Proceedings of the Fourth International Conference on Operational Research*, Boston, Mass., 1966; "The making of decision theory," P. C. Fishburn, pp. 369–388 in *Decision Science and Technology: Reflections on the Contributions of Ward Edwards*, J. Shanteau, B. Mellers, D. Schum, editors, Kluwer Academic Publishers, Boston, 1999; "Decision analysis: A personal account of how it got started and evolved," H. Raiffa, *Operations Research*, 50, 1, 2002, 179–185]

1966 Analysis of scheduling algorithms

The field of scheduling saw a revival in the late 1960s when researchers considered scheduling tasks on processors of computers rather than machines in a job shop. The search for approximate algorithms that could perform the scheduling rapidly led to the notion of guaranteed performance ratios, where the approximate solution is provably close to an optimal solution (say within a constant factor). Ronald Graham's work on multiprocessor scheduling provided a rich example on how to establish such guarantees and when they may prove to be elusive. This work and related research on bin packing led to the emergence of the analysis of heuristics as a very active area within both operations research and computer science in the 1970s. ["Bounds for certain multiprocessor anomalies," R. L. Graham, *Bell Systems Technical Journal*, 45, 1966, 1563–1581; "Bounds for multiprocessing timing anomalies," R. L. Graham, *SIAM Journal on Applied Mathematics*, 17, 1969, 263–269; *Computer and Job Shop Scheduling Theory*, E. G. Coffman, Jr., editor, John Wiley & Sons, New York, 1976]

1966 Criminal Justice: President's Crime Commission Science and Technology Task Force

As part of President Johnson's Commission on Law Enforcement and the Administration of Justice, a special task force on Science and Technology Task was formed under the direction of Alfred Blumstein (Institute for Defense Analysis). Other OR members of the Task Force included Saul I. Gass (IBM) and Richard C. Larson (MIT). The task force report and subsequent developments showed how OR methods can be used to analyze the systemic problems of the courts and the operational problems of law enforcement. Gass, Blumstein, and Larson were all presidents of ORSA (1976, 1977, and 1993, respectively), Blumstein was president of TIMS in 1987–1988, and Blumstein and Larson were presidents of INFORMS (1996 and 2005, respectively). [*Task Force Report: Science and Technology*, President's Commission on Law Enforcement and the Administration of Justice, Government Printing Office, Washington, DC, 1967; *Urban Operations Research*, R. C. Larson, A. R. Odoni, Prentice-Hall, Inc., Englewood Cliffs, 1981; "Crime modeling," A. Blumstein, *Operations Research*, 50, 1, 2002, 16–24]

Alfred Blumstein

1967 Games with incomplete information

How can game-theoretic models be extended to handle competitive situations when some players have incomplete information about some important parameters of the game such as payoff functions, other players' strategies, or information about the game available to other players? John C. Harsanyi answered this important question in a series of three papers in *Management Science*. He assumed that the information available can be modeled by Bayesian probability distributions on the parameters of interest. The first paper showed that for these Bayesian players, the game with incomplete information can be transformed into an equivalent Bayesian game with complete information (in the sense of having complete knowledge of the probability distribution governing the lottery). The second paper explored the correspondence between the Nash equilibrium for this Bayesian game and Bayesian equilibrium points of the original game. The third paper discusses the main properties of the related basic probability distribution from which the players' subjective probability distributions can be derived as conditional probability distributions. David Kreps and Ariel Rubinstein (1997) remark: "After Nash equilibrium, Harsanyi's definition of games with incomplete information is perhaps the single most important innovation from the point of view of modern economic applications." ["Games with incomplete information played by Bayesian players: Part I. The basic model," J. C. Harsanyi, *Management Science*, 14, 3, 1967, 159–182; "Part II: Bayesian equilibrium points," J. C. Harsanyi, *Management Science*, 14, 5, 1968, 320–334; "Part III. The basic probability distribution of the game," J. C. Harsanyi, *Management Science*, 14, 7, 1968, 486–502; "An appreciation," D. Kreps, A. Rubinstein, pp. xi–xv in *Classics in Game Theory*, H. W. Kuhn, editor, Princeton University Press, Princeton, 1997 (this book reprints the three Harsanyi papers)]

1967 *The Theory of Scheduling*, Richard W. Conway, William L. Maxwell, Louis W. Miller, Addison-Wesley, Reading (Dover reprint 2003)

This is the first book that provided a complete and systematic treatment of the theoretical aspects of scheduling. Based on a graduate OR course and written as a text, it brought the full range of techniques (algebraic, stochastic, simulation) for resolving job shop and other scheduling problems to the attention of the OR research, practitioner and academic communities.

1967 *Introduction to Operations Research*, Frederick S. Hillier, Gerald J. Lieberman, Holden-Day, Oakland

This is a widely-used introductory OR text. Aimed at junior and senior undergraduates and first-year graduates students, it was adopted both by business and engineering schools. Now in its seventh edition (2001, McGraw-Hill Book Co.), it is as popular as ever. Lieberman was president of TIMS in 1980–1981.

Frederick S. Hillier

Gerald J. Lieberman

1967 The computation of economic equilibria

Powerful and very general fixed point theorems guarantee the existence of an equilibrium solution, but one may ask: Is there an effective algorithm for computing the numerical solution of the neoclassical model of economic equilibrium? Herbert E. Scarf (1991) describes how he started to think about this question in 1963. He sought a constructive procedure for finding the equilibrium without appealing to the standard fixed point arguments. Drawing upon the earlier work of Carl Lemke on bimatrix games, Scarf (1967) introduced the notion of "primitive sets" and proved a combinatorial result that yielded an iterative algorithm for approximating the fixed point. Subsequently, this result was seen to be closely related to Sperner's lemma, a combinatorial result that has been used to prove Brouwer's fixed point theorem. Later work showed that the primitive sets used by Scarf can be replaced by a class of matrices found by Terje Hansen (1967) in the course of his doctoral studies, or alternatively, with a special simplicial subdivision of the simplex identified by Harold Kuhn (1968). The algorithmic procedures based on these two methods turned out to be identical and provided an effective computational procedure for finding fixed points. Scarf's 1973 book contains a detailed account of the conceptual development of his procedure for computing fixed points. ["The approximation of fixed points of a continuous mapping," H. E. Scarf, *SIAM Journal of Applied Mathematics*, 15, 5, 1967, 1328–1343; "On the approximation of a competitive equilibrium," Terje Hansen, Ph.D. dissertation, Yale University, 1968; "Simplicial approximation of fixed points," H. W. Kuhn, *Proceedings of the National Academy of Sciences*, 61, 1968, 1238–1242; *The Computation of Economic Equilibria*, Herbert Scarf (with Terje Hansen), Cowles Foundation for Research in Economics, Yale University Press, New Haven, 1982; "The origins of fixed point methods," Herbert E. Scarf, pp. 126–134 in *History of Mathematical Programming, A Collection of Personal Reminiscences*, J. K. Lenstra, A. H. G. Rinnooy Kan, A. Schrijver, editors, North-Holland, Amsterdam, 1991]

Herbert E. Scarf

1968 METRIC (Multi-Echelon Technique for Recoverable Item Control)

The METRIC model, developed by Craig C. Sherbrooke, evolved from the Air Force logistics program of the RAND Corporation. It was the first multi-echelon, multi-item computer-based inventory model proposed for implementation. Its military application involved repairable items of high value for which a one-for-one ordering policy is appropriate. The initial model could handle a single depot, several bases, and a large number of items. METRIC allows for three modes of operation: optimization for new procurement, evaluation of the existing distribution of stock, and redistribution of system stock between the bases and depot. Reviewing the development of METRIC within RAND, Murray Geisler stated that METRIC was well received in the Air Force and became a central policy focus for the Advance Logistics System (ALS). METRIC has been extended to multiple levels and richer settings. ["METRIC: A multi-echelon technique for recoverable item control," C. C. Sherbrooke, RM-078-PR, The RAND Corporation, Santa Monica, 1966, also in *Operations Research*, 16, 1, 1968, 122–141; *A Personal History of Logistics*, M. A. Geisler, Logistics Management Institute, Bethesda, Maryland, 1986; *Optimal Inventory Modeling of Systems: Multi-Echelon Techniques*, C. C. Sherbrooke, John Wiley & Sons, New York, 1992]

1968 Outranking procedures for multicriteria decision making (ELECTRE)

Bernard Roy

In multicriteria decision problems, an alternative i is said to outrank another alternative j if one can conclude that i is at least as good as j. This concept is imbedded into the ELECTRE methods developed by Bernard Roy. The results of an ELECTRE analysis is a ranking of the alternatives. ["Classement et choix en présence de points de vue multiples (La méthode ELECTRE)," B. Roy, *RIRO*, 8, 1968, 57–75; *Multicriteria Methodology for Decision Aiding*, B. Roy, Kluwer Academic Publishers, Boston, 1996]

1968 Decision Sciences Institute (DSI) founded

Originally founded as the American Institute for Decision Sciences, the Decision Sciences Institute is a multidisciplinary international association dedicated to advancing knowledge and improving instruction in all business and related disciplines. Dennis E. Grawoig was its first president.

1968 *Nonlinear Programming: Sequential Unconstrained Minimization Techniques*, Anthony V. Fiacco, Garth P. McCormick, John Wiley & Sons, New York

Recipient of the 1968 ORSA Lanchester prize for the best English language publication in OR, this book provided a unified theory (SUMT) on methods and computational procedures for transforming and solving a constrained minimization problem by a sequence of unconstrained minimizations of an appropriate auxiliary function. SUMT has also been shown to provide a basis for more recent work on interior point methods for solving linear programming problems.

1968 *Utility Theory: A Book of Readings*, Alfred N. Page, editor, John Wiley & Sons, New York

This volume is the first collection of articles and excerpts from books that relate primarily to the economic concept of utility. It includes Jeremy Bentham's "An Introduction to the Principles of Morals and Legislation" and, among others, works by R. G. D. Allen, Kenneth J. Arrow, Daniel Bernoulli, Milton Friedman and Leonard J. Savage, Vilfredo Pareto, Paul A. Samuelson, George J. Stigler, and John von Neumann and Oskar Morgenstern. It provides the first English translation of portions of Pareto's *Manuel d'Économie Politique* (*Ophélimité*), including the important discussion and appendix pertaining to Pareto optimality.

1968 *Decision Analysis: Introductory Lectures on Choices under Uncertainty*, Howard Raiffa, Addison-Wesley, Reading

In this series of "lectures," Raiffa presents his prescriptive Bayesian (subjective) approach for "how an individual who is faced with a problem of choice under uncertainty should go about choosing a course of action that is consistent with his personal basic judgments and preferences." In the final chapter, Raiffa discusses how decision analysis relates to game theory, operations research, and systems analysis. In Raiffa (2002), he notes that this book "documents the paradigmatic shift from statistical decision theory to decision analysis." ["Decision analysis: A personal account of how it got started and evolved," H. Raiffa, *Operations Research*, 50, 1, 2002, 179–185]

1968 *The Art of Computer Programming I: Fundamental Algorithms*, Donald Knuth, Addison-Wesley, Reading

Donald Knuth

Often called the bible of computer science, this work has had a profound organizing influence on the field. Volumes II and III, on seminumerical algorithms (1969) and sorting and searching (1973), are frequently cited by the OR community interested in algorithms. The 1974 Turing Award of the Association of Computer Machinery was awarded to Donald Knuth for the significant contributions that three volumes made to computer science. ["Donald Knuth," pp. 343–351 in *Portraits in Silicon*, R. Slater, MIT Press, Cambridge, Mass., 1987]

1969 New York City RAND Institute (NYCRI)

The RAND Corporation established the NYCRI to aid in the resolution of a variety of public policy problems. It helped to demonstrate how operations research and related analytical and computer-based methods could be of value to municipalities. The Institute's staff analyzed job training programs, nurse shortages, rent control, fire department management policies, and Jamaica Bay's water quality. [*Fire Department Deployment Analysis: A Public Policy Case Study*, W. E. Walker, J. M. Chaiken, E. J. Engels, editors, North-Holland, New York, 1979; "RAND Corporation," G. H. Fisher, W. E. Walker, pp. 690–695 in *Encyclopedia of Operations Research and Management Science*, 2nd edition, S. I. Gass, C. M. Harris, editors, Kluwer Academic Publishers, Boston, 2001]

1969 Advertising

The MEDIAC (Media Evaluation Using Dynamic and Interactive Applications of Computers) model, developed by John D. C. Little and Leonard M. Lodish, was one of the first marketing decision support systems. MEDIAC, formulated as a mathematical program, addressed media selection issues. The objective was to maximize total sales potential subject to constraints on current exposure value, media usage, and budgetary allocations. That is, it allocated a fixed budget over time and market segments. Its computer-based implementation was as an on-line conversational system. Due to the model's rather intractable nonlinear and integer conditions, a heuristic solution procedure was needed to make it operational. ["A media planning calculus," J. D. C. Little, L. M. Lodish, *Operations Research*, 17, 1, 1969, 1–35]

1969 First ARPANET/INTERNET site

In 1966, the Advanced Research Projects Agency (ARPA) recruited Lawrence G. Roberts from MIT to lead the development and installation of ARPA's proposed data network project. The specifications for what was termed the ARPANET were prepared in 1968 and a contract to implement and deploy it in 1969 went to Bolt, Beranek, and Newman, a computer consulting company. ARPA chose the University of California, Los Angeles (UCLA) to be the first host node to join the ARPANET. Leonard Kleinrock, a faculty member at UCLA and a consultant to ARPA, directed the UCLA effort. Kleinrock was responsible, among other things, for the packet switching concept of the ARPANET that, along with dynamic resource sharing, aids in making efficient use of the network's data transmission capacity. [*Inventing the Internet*, Janet Abbate, MIT Press, Cambridge, 1999; "Creating a mathematical theory of computer networks," Leonard Kleinrock, *Operations Research*, 50, 1, 2002, 125–131]

What hath God wrought?:

On September 2, 1969, messages began to move between the UCLA host computer and the co-located Interface Message Processor switch, a minicomputer. Stanford Research Institute joined a month later as the second ARPANET host node. The first host-to-host ARPANET message was sent from UCLA to Stanford on October 29, 1969.

Leonard Kleinrock

1969 ***Principles of Operations Research***, Harvey M. Wagner, Prentice-Hall, Englewood Cliffs

Written as an undergraduate and graduate text for students in business, economics, engineering, liberal arts and public administration, this book set a new standard for such texts in terms of its inclusiveness and clarity of writing. It received ORSA's Lanchester prize for the best publication in the English language, as well as the AIIE Book Award. Wagner was president of TIMS in 1973–1974.

How to drive an algorithm:

In explaining why he is persuaded that executives and managers must understand the principles of OR methodologies, Wagner (1969, pp. 91–92) offered the following analogy. "In learning to drive a car, it strains the intellect only mildly to learn to speed up, slow down, go forward or backward, turn, and so forth – in short, to master the skill of driving. To qualify as a topnotch driver, however, you need to know more. You should have an idea of how to care for the battery . . . a knowledge of the braking mechanism . . . the radiator's function. . . . you can obviously be an excellent driver without the training of an automobile mechanic." Wagner sums up: If managers want to maintain control, they must nurture their insight to the approach. It only takes a modest effort; you do not need to be an expert theoretician.

Harvey M. Wagner

1969 *System Simulation*, Geoffrey Gordon, Prentice-Hall, Englewood Cliffs

Written by the developer of GPSS, the first special-purpose computer-based language for discrete simulation, this book gives an overview of six different simulation languages, and discusses the broader issues of system and model development as they relate to simulation issues.

1969 *Network Flow, Transportation and Scheduling*, Masao Iri, Academic Press, New York

This book gives a self-contained exposition of network-flow problems and related solution algorithms. Much of the exposition is based on Iri's pioneering research in network flows and related algorithms.

1969 *Integer Programming and Network Flows*, Te Chiang Hu, Addison-Wesley, Reading

Te Chiang (TC) Hu

A full and basic treatment that was the first to describe the standard relationships between linear programming, network analysis and integer programming, as well as advanced topics and algorithms.

1969 *The Single Server Queue*, Jacob W. Cohen, North-Holland, Amsterdam (Revised edition, 1982)

This comprehensive treatment of the single-server queue has long served as an advanced and highly mathematical treatment of queues. The book makes heavy use of analytical machinery and is one of the few sources to cover the transient solution of the $M/M/1$ in full. The 1982 edition of the book runs over 800 pages in length.

1970 *Interfaces* begins publication

The main objective of this journal, sponsored by TIMS, was the publishing of papers dealing with operational problems using management science. It was first published under the title, *The Bulletin*. Leonard S. Simon was its first editor.

1970 Lagrangian relaxation for discrete optimization

Lagrangian relaxation is a method for obtaining bounds and solutions for integer-programming problems. It was first developed by Michael Held and Richard M. Karp. The method assumes that the constraints of the original integer-programming problem can be separated into two parts, one of which has a special structure that allows the problem to be solved easily. The complicating constraints are then relaxed and the problem is solved retaining only the easy subset of constraints. In his survey, Arthur M. Geoffrion gave the name Lagrangean [sic] relaxation to the process. Geoffrion also defined the integrality property that clarifies the relation between the Lagrangian relaxation of an integer program with its usual linear programming relaxation that drops the integrality conditions. ["Three problems in capital rationing," J. H. Lorie, L. J. Savage, *Journal of Business*, 28, 1955, 229–239; "Generalized Lagrange multiplier method for solving problems of optimum allocation of resources," H. Everett, III, *Operations Research*, 11, 1963, 399–417; "The traveling salesman problem and minimum spanning trees," M. Held, R. M. Karp, *Operations Research*, 18, 6, 1970, 1138–1162; "The traveling salesman problem and minimum spanning trees: Part II," M. Held, R. M. Karp, *Mathematical Programming*, 1, 1, 1971, 6–25; "Lagrangean relaxation for integer programming," A. M. Geoffrion, *Mathematical Programming Study*, 2, 1974, 82–114; "The Lagrangian relaxation method for solving integer programming problems," M. L. Fisher, *Management Science*, 27, 1, 1981, 1–18; *The Theory of Linear and Integer Programming*, A. Schrijver, John Wiley & Sons, New York, 1986]

Langrangian evolution:

According to Fisher (1981): "There were a number of forays prior to 1970 into the use of Lagrangian methods in discrete optimization, including the Lorie–Savage (1955) approach to capital budgeting, Everett's (1963) proposal for generalizing Lagrange multipliers, and the philosophically related device of generating columns by solving an easy combinatorial problem when pricing out in the simplex method. However, the 'birth' of the Lagrangian approach as it exists today occurred in 1970 when Held and Karp used a Lagrangian problem based on minimum spanning trees to devise a dramatically successful algorithm for the traveling salesman problem."

Arthur M. Geoffrion

1970 Ellipsoid method

The standard simplex method has been shown to be an exponential-time algorithm in that certain linear-programming problems (e.g., the Klee–Minty problems) require the simplex method to find all (exponentially large) solutions. Thus, the simplex method is not a polynomial-time algorithm (efficient algorithm). The ellipsoid method was developed by Naum Z. Shor for solving nonlinear convex programming problems, with further extensions by David B. Yudin and Arkadi S. Nemirovski. Leonid G. Khachian adapted the ellipsoid method to solve linear-programming problems and showed that it can find an optimal solution within a polynomial-bounded number of iterations. Although of theoretical importance, the ellipsoid algorithm exhibits poor computational properties in terms of the time and the number of iterations required to solve problems of reasonable size. Thus, it is not used in practice. ["Utilization of the operation of space dilation in the minimization of convex functions," N. Z. Shor, *Kibernetika*, 1, 1970, 6–12 (Russian), translated in *Cybernetics*, 6, 1970, 7–15; "Evaluation of the informational complexity of mathematical programming problems," D. B. Yudin, A. S. Nemirovski, *Èkonomika i Mathematicheskie Metody*, 12, 1976, 128–142 (Russian), translated in *Matekon*, 13, 2, 1976–1977, 3–25; "A polynomial algorithm for linear programming," L. G. Khachian, *Doklady Akademii Nauk SSSR*, 244, 1979, 1093–1096 (Russian), translated in *Soviet Mathematics Doklady*, 20, 1979, 191–194; *Theory of Linear and Integer Programming*, A. Schrijver, John Wiley & Sons, New York, 1986]

Arkadi S. Nemirovski

George B. Dantzig   Leonid G. Khachian

1970 Efficient network flow algorithms

The original Ford–Fulkerson algorithm for solving the maximum flow problem (between a source node s and sink node t of a network) is not a polynomial-time algorithm. In 1969–1970, Jack Edmonds and Richard M. Karp and, independently, the Russian mathematician Efim A. Dinits developed the first efficient algorithms for the (s, t) maximum flow problem. For a network with m edges and n nodes, the running times of the algorithms are $O(nm^2)$ and $O(n^2m)$, respectively. The two efforts had certain themes in common. For example, both used the concept of a flow-augmenting path with the smallest number of edges. Dinits's algorithm became known to the West in the late the 1970s. In fact, it took some time before researchers came to view his algorithm as an efficient implementation of Ford and Fulkerson's procedure. The design of efficient network algorithms gained momentum in the 1980s with the work of Robert E. Tarjan, and Ravindra K. Ahuja and James B. Orlin. ["Algorithm for solution of a problem of maximum flow in a network with power estimation," E. A. Dinits, *Soviet Mathematics Doklady*, 11, 1970, 1277–1280; "Theoretical improvements in algorithmic efficiency for network flow problems," J. Edmunds, R. M. Karp, *Journal of the ACM*, 19, 1972, 248–264; *Data Structures and Network Algorithms*, R. E. Tarjan, SIAM, Philadelphia; "Design (with analysis) of efficient algorithms," D. Gusfield, Chapter 8 of *Handbooks in Operations Research and Management Science, Vol. 3: Com-*

puting, E. G. Coffman, Jr., J. K. Lenstra, A. H. G. Rinnooy Kan, editors, North-Holland, New York, 1992; *Network Flows*, R. K. Ahuja, T. L. Magnanti, J. B. Orlin, Prentice-Hall, Englewood Cliffs, 1993]

1970 Multiple Criteria decision making (MCDM)

Mulitiple criteria decision making (MCDM) or multiple-criteria decision analysis (MCDA) are concerned with the selection of the "best" alternative from a given set of alternatives, based on how the alternatives are evaluated against a set of criteria or objectives. The resolution of such decision problems has been a major OR research area since its inception. Attempts to resolve them, however, have a long history: utility theory (J. Bentham, V. Pareto, J. von Neumann and O. Morgenstern), vector maximization (H. Kuhn and A. W. Tucker), measure of value (C. W. Churchman and R. Ackoff). Its place here on the *Annotated Timeline* can be traced to two important conferences: (1) the 7th Mathematical Programming Symposium held in The Hague in 1970, and (2) the 1972 Multiple Criteria Decision Making Conference held at the University of South Carolina. Many of the Symposium's papers, especially those dealing with multiple criteria, were published in *Mathematical Programming*, Vol. 1, Nos. 1, 2, and 3, 1971, while those given at the Conference are in the proceedings *Multiple Criteria Decision Making*, J. L. Cochrane, M. Zeleny, editors, University of South Carolina Press, Columbia, 1973. Related research and applied developments, some earlier and some later, include the work of the following: R. Benayoun, J. de Montgolfier, J. Tergny, O. Larichev (STEM); W. Edwards (SMART); A. M. Geoffrion, J. S. Dyer, A. Feinberg (Interactive multicriteria optimization); S. Zionts, J. Wallenius (Interactive methods for multiple criteria problems); R. Steuer (Interactive weighted Chebyshev procedure for multiple-objective programming); B. Roy (Outranking/ELECTRE methods); M. Zeleny (MCDM); J. P. Brans, B. Mareschal, P. Vincke (PROMETHEE); T. L. Saaty (Analytic Hierarchy Process); A. P. Wierzbicki (reference point methods). [*Making Multiple-Objective Decisions*, M. Mollaghasemi, J. Pet-Edwards, IEEE Computer Society Press, Los Alamitos, 1997; *Multiple Criteria Decision Analysis*, V. Belton, T. J. Stewart, Kluwer Academic Publishers, Boston, 2002]

1970 *Time Series Analysis*, George E. P. Box, Gwilym M. Jenkins, Holden-Day, San Francisco

Considered the gold standard in forecasting time series, the class of autoregressive integrated moving averages (ARIMA) models of Box and Jenkins make extensive use of the statistical properties of the series. In the class of ARIMA(p, d, q) models, where p, d, and q are integers parameters chosen by the modeler, the resulting forecasts are expressed as a sum of weighted p autoregressive and q moving averages terms whose coefficients must be estimated using least-squares techniques. The standard reference is the text by Box and Jenkins, although the fundamental concepts date back to at least 1962. ["Some statistical aspects of adaptive optimization and control," G. E. P. Box, G. M. Jenkins, *Journal of the Royal Statistical Society*, B, 24, 1962, 297ff; "Forecasting," R. G. Brown, pp. 3–26 in *Handbook of Operations Research: Models and Applications*, Vol. 2, J. J. Moder, S. E. Elmaghraby, editors, 1978; "A conversation with George Box," Morris De Groot, *Statistical Science*, 2, 3, 1987, 239–258]

The time-dependent chase:

In De Groot (1987), Box recounts that the origin of his interest in times-series forecasting was an industrial problem in which the yield of a chemical reactor changed over time. The question was how to pursue the temperature that gave the maximum yield. Box and Jenkins started with this optimization problem and realized it was a control problem. But the control problem required forecasting for nonstationary time series, which led to the Box–Jenkins model.

George E. P. Box

1970 *Convex Analysis*, R. Tyrrell Rockafellar, Princeton University Press, Princeton

Based on the author's classroom notes, this book presents a modern treatment of convexity as applied to mathematical extremum problems. Material covered includes: systems of inequalities, the minimum and maximum of a convex function over a convex set, Lagrange multipliers, and minimax theorems, as well as basic material on convex sets and functions.

Convex lineage:

As Rockafellar notes, many aspects of his exposition are due to Werner Fenchel, who, as a visiting professor (1951), Department of Mathematics, Princeton University, taught a course in "Convex Cones, Sets, and Functions." A mimeographed publication with that title, authored by Fenchel, was published in 1953 by Princeton University from notes of Fenchel's lectures taken by Donald W. Blackett. Fenchel's lectures covered properties of convex cones, convex sets, and convex functions in finite dimensional spaces, with applications to the theory of games and convex programming problems.

R. Tyrrell Rockafellar

1970 *Convexity and Optimization in Finite Dimensions I*, Josef Stoer and Christoph Witzgall, Springer-Verlag, New York

This book provides the necessary theoretical background for the "arithmetic" of the expanding field of convex optimization: linear programming, algebra of linear inequalities, geometry of polyhedra, the topology of convex sets, and the analysis of convex functions. It emphasizes linear and convex duality theory and its impact on modern optimization techniques.

Josef Stoer

Christoph Witzgall

1971 The class NP and NP-complete problems

Stephen A. Cook's historic 1971 paper proved that every set of strings accepted in polynomial time by a nondeterministic Turing machine is reducible to the propositional satisfiability problem (SAT) in polynomial time. This showed that P = NP if and only if SAT is solvable in polynomial time. Thus SAT was identified as the archetypically intractable problem unless P = NP. Richard M. Karp immediately recognized the importance and consequence of this result. He identified a class of problems he called polynomial complete (now called NP-complete) that were also archetypically intractable in Cook's sense. Karp's celebrated 1972 paper provided a list of such problems, which reduced a number of classical combinatorial optimization problems to SAT. Cook and Karp received the ACM Turing Award in 1982 and 1985, respectively. ["The complexity of theorem-proving procedures," Stephen A. Cook, pp. 151–158 in *Proceedings of the 3rd Annual ACM Symposium on the Theory of Computing Machinery*, 1971; "Reducibility among combinatorial problems," Richard M. Karp, pp. 85–103 in *Complexity of Computer Computations*, R. E. Miller, J. W. Thatcher, editors, Plenum, New York, 1972; "An overview of computational complexity," S. A. Cook, pp. 411–431 in *ACM Turing Award: The First Twenty Years Lectures 1966–1985*, ACM Press, New York, 1987; "Combinatorics, complexity, and randomness," R. M. Karp, pp. 433–453 in *ACM Turing Award: The First Twenty Years Lectures 1966–1985*,

ACM Press, New York, 1987; "The mysteries of algorithms," R. Karp, pp. 147–162 in *People and Ideas in Theoretical Computer Science*, C. S. Calude, editor, Springer-Verlag, Singapore, 1999]

Stephen A. Cook

Richard M. Karp

On getting the word out:

Karp (1999) recalls how his ideas were first made public: "It was an exciting time because I had a clear conviction that I was doing work of great importance. Most of the reductions I was after came easily, but the NP-Completeness of the Hamiltonian circuit problem eluded me, and the first proofs were given by Gene Lawler and Bob Tarjan, who were among the first to grasp the significance of what I was doing. The first opportunity to speak about NP-completeness came at a seminar at Don Knuth's home. In April 1972, I presented my results before a large audience at a symposium at IBM and in the following months I visited several universities to give talks about NP-Completeness."

1971 Decision support system

Michael S. Scott Morton (1971) first proposed the concept of a Management Decision System (MDS) based on his observations of middle-level and senior managers who used a system to support their decisions. He defined MDS as: "... an approach involving the analysis of key decisions and the design of support for these decisions ... (involving) the use of interactive graphics terminals, a multiple-access computer, and a data-bank and model-bank relevant to the problem." The term MDS was soon changed to Decision Support System (DSS), for which the 1973 text by Peter G. W. Keen and Morton served as the standard reference. [*Management Decision Systems: Computer-Based Support for Decision-Making*, M. S. Scott Morton, Division of Research, Graduate School of Business Administration, Harvard University, Boston, 1971; *Decision Support Systems*, P. G. W. Keen, M. S. Scott Morton, Addison-Wesley, Reading, 1973]

1971 The Mathematical Programming Society (MPS)

This international society encourages theoretical research, applications and computational developments of all aspects of constrained optimization. It sponsors the triennial International Symposium on Mathematical Programming. George B. Dantzig was the first Chairman of the Society. ["Mathematical programming: Journal, society, recollections," M. L. Balinski, pp. 5–18 in *History of Mathematical Programming*, J. K. Lenstra, A. H. G. Rinnooy Kan, A. Schrijver, editors, North-Holland, Amsterdam, 1991]

1971 *Networks* began publication

In 1970, Howard Frank, Ivan T. Frisch, and Richard Van Slyke formed the Network Analysis Corporation, a consulting company with expertise in solving network problems. The company became a center for researchers interested in network applications. The idea of having a journal dedicated to networks first occurred to Frank, Frisch, and David Rosenbaum. Frisch served as editor-in-chief until 1978, along with Frank T. Boesch and Daniel J. Kleitman as editors. Citing a number of different applications areas (computer networks, air and road traffic, and pipelines), the mission of the journal was to "unify the materials relevant to the study of these problems and focus the attention of researchers and designers on the central theme of networks." ["The early days of *Networks*," I. T. Frisch, *Networks*, 37, 1, 2001, 1–7]

1971 *World Dynamics*, Jay W. Forrester, MIT Press, Cambridge

This book extends Forrester's systems dynamics model to the systemic problems faced by the world in terms of population growth, energy, food, environmental and related concerns. It describes the *World3* model that was central to the 1972 *Limits of Growth* study. The impending crises predicted by the model to occur by the end of the 20th century attracted much attention and engendered controversy in such journals as *Nature* and *Science*. [*On Systems Analysis: An Essay Concerning the Limitations of Some Mathematical Methods in the Social, Political, and Biological Sciences*, David Berlinksi, MIT Press, Cambridge, 1976]

1971 *Great Expectations: The Theory of Optimal Stopping*, Yuan Shih Chow, Herbert Robbins, David Seigmund, Houghton-Mifflin Company, Boston (Dover reprint 1991)

This was the first book published in the United States devoted entirely to the topic of optimal stopping. In certain operations research problems involving sequential decisions, the decision maker has the choice between stopping or continuing to observe the process. If $\{Y_n,\ n = 1, 2, \ldots\}$ denotes the sequential random variables in discrete time and stopping at stage n earns a reward of $X_n = f_n(Y_1, \ldots, Y_n)$, the decision maker seeks an optimal stopping rule that would maximize return. A stopping rule T is an integer-valued random variable such that the decision to stop at time n $(T = n)$, is made on the basis of the past values of $Y_1, \ldots, Y_n$ alone. Although there was a good deal of previous research on optimal stopping, the authors' work, using martingale theory, developed the subject significantly. ["Optimum character of the sequential probability ratio test," A. Wald, J. Wolfowitz, *Annals of Mathematical Statistics*, 19, 1948, 326–339; "Bayes and minimax solutions of sequential decision problems," K. J. Arrow, D. Blackwell, M. A. Girshick, *Econometrica*, 17, 1949, 213–243; "On optimal stopping rules," Y. S. Chow, H. Robbins, *Z. Wahrscheinlichkeitstheorie*, 2, 1963, 33–49; "A class of optimal stopping time problems," Y. S. Chow, H. Robbins, pp. 419–426 in *Proceedings of the Fifth Berkeley Symposium on Mathematical Statistics and Probability*, Vol. 1, University of California Press, Berkeley, 1967]

1972 International Institute for Applied Systems Analysis (IIASA)

The International Institute for Applied Systems Analysis is a nongovernmental research institute located in Laxenburg, Austria. It was founded by academies of science or equivalent institutions of twelve nations. Howard Raiffa was its first director. The original motivation for its establishment was to enable scientists from East and West to work together on problems of common concern. The scientific staff is now drawn from all countries of the world. IIASA has been instrumental in the development of global models for analyzing policy issues of the environment, energy and other resources, economic, and population.

1972 Soft systems methodologies

Soft systems methodology (SSM), or soft OR, applies and extends the ideas and concepts of (hard) OR to real-world problems often referred to as "messes" or "wicked." That is, SSM does not assume a systemic view can be imposed on such problems, but the ideas of systems analysis help form the process of inquiry. First proposed by Peter Checkland, SSM was developed as a means of resolving unstructured management, planning, and public policy situations which involve multi-objectives that are often unclear or contradictory. ["Towards a systems-based methodology for real-world problem solving," P. Checkland, *Journal of Systems Engineering*, 3, 2, 1972, 87–116; *Systems Thinking, Systems Practice*, P. Checkland, Wiley, Chichester, 1981; "Soft systems methodology," P. Checkland, pp. 766–770 in *Encyclopedia of Operations Research and Management Science*, 2nd edition, S. I. Gass, C. M. Harris, editors, Kluwer Academic Publishers, Boston, 2001]

1972 Franz Edelman Award for Management Science Practice established

The College of Practice of TIMS established the Edelman Award to recognize outstanding examples of management science in practice. This is a competition in which organizations submit their accomplishments to a set of evaluative judges who then select a subset of the applicants to participate in final run-off presentations. The first award (1972) went to the Pillsbury Company. The award is now sponsored jointly by INFORMS and the INFORM's College of Practice of Management Science (CPMS).

How do you get to Carnegie Hall?:

Franz Edelman established the Operations Research Group at RCA, one of the earliest industrial OR/MS groups in North America. He worked for over 30 years at RCA and is considered a pioneer of innovation and application of management science. He believed in practice.

Franz Edelman
1922–1982

1972 The simplex method is not a polynomial-time algorithm

In their paper, "How good is the simplex method," Victor Klee and George J. Minty provided an example of a linear-programming problem for which the simplex method would have to evaluate all vertices of the defining polytope, where the number of vertices can be made to grow exponentially. Such "worst-case" problems require $(2^d - 1)$ iterations for a linear program with d variables and $m = 2d$ inequalities. ["How good is the simplex method," Victor Klee, George J. Minty, pp. 159–175 in *Inequalities – III*, O. Shisha, editor, Academic Press, New York, 1972]

The Klee–Minty problem:

Minimize $-x_d$
subject to
$x_1 \geqslant 0$
$x_1 \leqslant 1$
$-\varepsilon x_1 + x_2 \geqslant 0$
$\varepsilon x_1 + x_2 \leqslant 1$
....
$-\varepsilon x_{d-1} + x_d \geqslant 0$
$\varepsilon x_{d-1} + x_d \leqslant 1 \quad (x_j \geqslant 0; 0 < \varepsilon < \frac{1}{2})$

1972 *Limits to Growth*, Donella H. Meadows, Dennis L. Meadows, Jørgen Randers, W. W. Behrens III, Signet Books, Washington

The Club of Rome, an international group of industrialists, scientists, educators and others, commissioned a study of the environmental, energy, population, economic, agriculture and related concerns in terms of how they would impact the future of the world. This book presents the results of the study that were based on Jay W. Forrester's systems dynamics model, *World3*. The dire predictions of the study were quite controversial, both in terms of their interpretation and their genesis from a high level computer-based model. [*World Dynamics*, J. W. Forrester, MIT Press, Cambridge, 1971; *Models in the Policy Process*, M. Greenberger, M. A. Crenson, B. L. Crissey, Russell Sage Foundation, New York, 1976]

1972 *Urban Police Patrol Analysis*, Richard C. Larson, MIT Press, Cambridge

This book, an extension of the author's 1967 doctoral dissertation, describes the use of OR methods to plan the operations of police patrols. The topics include the analysis of travel times (for a dispatched unit to get to the scene of an incident), the allocation of patrol across different commands of a city, and the evaluation of the impact of automatic car locator technologies. The book, for which Larson received the 1972 Lanchester prize, reflects his experience with the police departments of Boston and New York.

1972 *Integer Programming*, Robert S. Garfinkel, George L. Nemhauser, John Wiley & Sons, New York

This text provided a comprehensive treatment of the first two decades of research on integer programming. It rapidly became a favorite text for graduate courses. Of particular interest are the historical notes at the end of each chapter which review the literature to 1972.

1972 *Human Problem-Solving*, Allen Newell, Herbert A. Simon, Prentice-Hall, Englewood Cliffs

This book is based on the authors' decade (1950s) of collaborative work in artificial intelligence and cognitive psychology. They believed that thinking can be explained in terms of two key abilities: (a) searching and (b) storing and retrieving fragments of knowledge. Knowledge was stored as production (IF ... THEN) rules, and the role of production rules in problem-solving was a key theme of the book. The 920-page tome, which took 14 years to write, introduces information processing, computer simulation, and problem-solving by heuristic search. It describes a theory of problem-solving the authors inferred from their empirical work. Edward Feigenbaum described the book as "perhaps the most important book on the scientific study of human thinking in the 20th century." Newell and Simon were awarded the 1975 Turing Award, marking the first time the award had joint recipients. [*Models of My Life*, Herbert A. Simon, Basic Books, New York, 1991; "Retrospective: Herbert Simon, 1916–2001," E. A. Feigenbaum, *Science*, 291, 5511, 2001, 2107]

1973 Options pricing

A month after the establishment of the Chicago Board Options Exchange, a research paper published by Fischer Black and Myron Scholes provided a formula for pricing options, which are claims on underlying financial instruments, including stock shares and foreign exchange. The pricing of an option is based on the same principles as the equilibrium market pricing for tangible assets: the efficient price is the one that fairly compensates the seller for accepting the obligation described by the options contract. This compensation depends on four factors: the term of the option, its exercise or strike price, the interest rate available on alternative investments, and the current price and future volatility of the underlying asset. Drawing on their training in mathematics, Black and Scholes were able to formulate the pricing problem as an application of geometric Brownian motion. They arrived at the now-famous Black–Scholes pricing formula in 1969, but the full paper was delayed until 1973. A paper by Robert C. Merton solved the same problem almost simultaneously. Scholes and Merton were awarded the 1997 Nobel prize in economics for a new method to determine the value of derivatives. Black, who died in 1995, was thus not eligible for the 1997 Nobel prize. The Black–Scholes–Merton formulations laid the foundation for the area of financial engineering and paved the way for the subsequent development of huge markets for financial derivatives. ["The pricing of options and corporate liabilities," F. Black, M. Scholes, *Journal of Political Economy*, 81, 1973, 637–659; "Theory of rational option pricing," R. C. Merton, *Bell Journal of Economics and Management Science*, 4, 1973, 141–183; "How we came up with the option formula," F. Black, *Journal of Portfolio Management*, 15, 1989, 4–8; *The Nobel Laureates: How the World's Greatest Economic Minds Shaped Modern Thought*, Marilu Hurt McCarthy, McGraw-Hill, New York, 2001, 272–280]

(©Nobel Foundation)
Robert C. Merton

(©Nobel Foundation)
Myron Scholes

1973 Near-optimal bin-packing algorithms

The bin-packing problem involves the packing of a set of weighted items into the minimum number of bins of unit capacity. While the bin packing problem is NP-complete, the research for near-optimal approximation algorithms constitutes one of the celebrated examples of the power of analysis of heuristics for combinatorial problems. An example of a simple heuristic is the *First Fit Algorithm* which places the next item in the first bin where it can fit, or in a new bin if none of the existing ones can accommodate it. The first significant result was due to Jeffrey Ullman who proved that the *First Fit Algorithm* requires at most $[(17/10)\mathrm{OPT} + 3]$ bins, where OPT is the minimum number of bins required. In his doctoral dissertation, David Johnson carried out an extensive analysis of bin packing algorithms including the *First Fit Decreasing Algorithm* (*FFD*), which first arranges the items in decreasing order of size and then applies the First Fit algorithm to this list. In an ingenious proof that runs over 100 pages in his dissertation, Johnson established a worst-case bound of $11/9$ for FFD, so that this algorithm is guaranteed never to be more than approximately 22% worse than optimal. Ronald L. Graham (1976) provides an overview of this body of research up to 1976, and its subsequent development is surveyed by E. G. Coffman, M. R. Garey, and D. S. Johnson (1997). ["The performance of a memory allocation algorithm," J. D. Ullman, Technical Report 100, Electrical Engineering Department, Princeton University, Princeton, 1971; *Near-Optimal Bin Packing Algorithms*, D. S. Johnson, Ph.D. dissertation, Department of Mathematics, MIT, Cambridge, 1973; "Bounds on the performance of scheduling algorithms," R. L. Graham, pp. 165–227 in *Computer and Job-Shop Scheduling Theory*, E. G. Coffman, Jr., editor, John Wiley & Sons, New York, 1976; "Approximation algorithms for bin-packing – a survey," E. G. Coffman, Jr., M. R. Garey, D. S. Johnson, pp. 46–93 in *Approximation Algorithms for NP-Hard Problems*, D. S. Hochbaum, editor, PWS, Boston, 1997]

1974 *Computers and Operations Research* begins publication

The international journal, *Computers and Operations Research*, was founded to emphasize "new and interesting applications of OR to problems of world concern and general interest." It was one in a series of such "Computers and ..." journals published by Pergamon Press. Samuel J. Raff served as the journal's editor from its inception through 2002.

1974 Hypercube queueing model

Rapid response of emergency services such as police cars and ambulances is a function of their location and number deployed in the field. Related issues are the design of patrol beats, the location of emergency service facilities, and the evaluation of dispatch policies. Building on his initial research for the President's Crime Commission Science and Technology Task Force (1966–1967), Richard C. Larson developed the hypercube queueing model that helps to answer the operational and planning concerns related to police dispatch and emergency response. The model has been installed and used by a number of police departments. ["A hypercube queueing model for facility location and redistricting in urban emergency services," R. C. Larson, *Computers and Operations Research*, 1, 1, 1974, 67–95; "Hypercube queueing model," R. C. Larson, pp. 373–377 in *Encyclopedia of Operations Research and Management Science*, 2nd edition, S. I. Gass, C. M. Harris, editors, Kluwer Academic Publishers, Boston, 2001]

Car 54, where are you?

Larson (2001) notes that the hypercube model's roots started with his work on the Crime Commission and with MIT-affiliated work with the Boston Police Department. From numerous hours riding around in the rear seats of police cars and standing behind police radio dispatchers, he learned that the fleet of police cars in an area of the city can be viewed as "spatially distributed servers" in a queueing system. "Customer inputs" to this queueing system are generated by citizens calling "911" and asking for emergency service.

Richard C. Larson

1974 GASP simulation languages

GASP is a flexible FORTRAN-based simulation language using event-scheduling control developed by Alan Pritsker and his co-workers. In the transition from GASP II to GASP IV, Pritsker and his students modified the notion of "event" and worked out the necessary changes to allow combined continuous and discrete modeling. Pritsker and C. Dennis Pegden designed SLAM as an extension of GASP IV by adding a network modeling capability reminiscent of GPSS. The network representation allows the user to visually structure the system as a network through which the entities flow. [*Simulation with GASP II: A FORTRAN Based Simulation Language*, A. B. Pritsker, P. J. Kiviat, Prentice-Hall, Englewood Cliffs, NJ, 1969; *The GASP IV Simulation Language*, A. B. Pritsker, John Wiley & Sons, New York, 1974; *Introduction to Simulation and SLAM*, A. Pritsker, C. D. Pegden, John Wiley & Sons, New York, 1979; "Alan Pritsker's multifaceted career: Theory, practice, education, entrepreneurship, and service," J. R. Wilson, D. Goldsman, *IIE Transactions*, 33, 2001, 139–147]

1974 OR established at Federal Express Corporation

Soon after it officially began operations in 1973, Federal Express established an OR Department that reported directly to Frederick W. Smith, Chairman and CEO. As noted by Smith: the OR Department played a role in the development of Federal Express's long-range plans and was instrumental in its becoming the world's largest air carrier; all major system changes, such as number and location of hubs and fleet composition analysis, were first modeled by the OR analysts several years in advance of the actual system change. ["Eyes on the Prize," P. Horner, *OR/MS Today*, 4, 1991, 34–38]

1974 First joint ORSA and TIMS meeting

The first joint national meeting of ORSA and TIMS was held on April 22–24, 1974, in Boston. This initiated the series of two yearly joint national meetings, the TIMS/ORSA Spring meeting and the ORSA/TIMS Fall meeting.

1974 *OR/MS Today* begins publication

OR/MS Today was the first joint publication of ORSA and TIMS.

1974 *Interfaces* becomes joint publication of TIMS and ORSA

As noted in its editorial statement and policy, *Interfaces* seeks to improve communication between OR/MS managers and professionals by publishing papers that describe practice and implementation of OR/MS in commerce, industry, government, or education.

1974 *A Guide to Models in Governmental Planning and Operations*, Saul I. Gass, Roger Sisson, editors, Environmental Protection Agency, Washington (also published by Sauger Books, Potomac, 1975)

This book, sponsored by the Environmental Protection Agency, contains chapters describing pioneering models applied to air pollution, water resources, solid waste management, urban development, transportation, housing, law enforcement and criminal justice, education, energy, health and policy analysis.

1974 *Fundamentals of Queueing Theory*, Donald Gross, Carl M. Harris, John Wiley & Sons, New York (third edition 1997)

This introduction to queueing theory has been popular for its clear and detailed exposition. Close to half the book is devoted to Markovian queueing models, and about a fifth is devoted to statistical inference and simulation. Of particular interest is the chapter detailing a case study in queueing by Georges Brigham, which addressed the optimal staffing of tool cribs with clerks in a Boeing plant. Gross and Harris were presidents of ORSA in 1989 and 1990, respectively. ["On a congestion problem in an aircraft industry," G. Brigham, *Operations Research*, 3, 4, 1955, 412–428]

Donald Gross

Carl M. Harris
1940–2000

1975 Probabilistic analysis of combinatorial algorithms

In 1975, Richard M. Karp committed himself to an investigation of probabilistic analysis of combinatorial algorithms. Of this decision, he writes: "I must say that this decision required some courage, since this line of research had its detractors .... I felt, however, that in the case of NP-complete problems we weren't going to get worst-case guarantees we wanted, and that the probabilistic approach was the best way and perhaps the only way to understand why heuristic combinatorial algorithms worked so well in practice." The first fruit of this labor was Karp's partitioning algorithm for the traveling salesman problem in a plane. Karp's work had the desired impact of launching the probabilistic analysis of algorithms as an area of research. The subsequent work of others on the probabilistic analysis of the simplex method constituted another triumph for this research program. Karp was awarded the 1977 Lanchester prize for his traveling salesman paper. ["The probabilistic analysis of some combinatorial search algorithms," R. M. Karp, pp. 1–19 in *Algorithms and Complexity: New Directions and Recent Results*, J. F. Traub, editor, Academic Press, New York, 1976; "Probabilistic analysis of partitioning algorithms for the traveling salesman problem in the plane," R. M. Karp, *Mathematics of Operations Research*, 2, 1977, 209–224; "The average number of pivot steps required by the simplex method is polynomial," K.-H. Borgwardt, *Zeitschrift für Operations Research*, 26, 1982, 157–177; "Combinatorics, complexity, and randomness," R. M. Karp, pp. 433–453 in *ACM Turing Award: The First Twenty Years Lectures 1966–1985*, ACM Press, New York, 1987]

1975 Genetic algorithms

A genetic algorithm is a heuristic procedure in which the randomized search mimics the mechanisms of natural selection. John H. Holland first developed such procedures in 1962 when he investigated the evolution of complex adaptive systems characterized by interacting genes. Holland's decade of computational research on the subject culminated in his 1975 book. The application of genetic algorithms to combinatorial optimization has grown steadily since the mid 1980s. To apply this method, one must first construct a representation of solutions as binary strings analogous to chromosomes. Representing each solution as a string, a population of solutions is allowed to evolve through successive generations, subject to a preset maximum on the number of solutions in each generation. Each new generation is obtained in three steps: First, the quality or fitness of each solution is evaluated. Second, the selection step identifies the solutions allowed to generate offspring using a probabilistic rule based on the relative fitness values of current solutions. Finally, new solutions are created by modifying a single solution (mutation) or combining features of a pair of solutions (crossover). The evolutionary principle of the "survival of the fittest" will then produce high-quality solutions to the original optimization problem. [*Adaptation in Natural and Artificial Systems*, J. Holland, The University of Michigan Press, Ann Arbor, 1975; *Complexity*, M. M. Wardrop, Simon & Schuster, New York, 1992; *An Introduction to Genetic Algorithms (Complex Adaptive Systems)*, M. Mitchell, MIT Press, Cambridge; "Evolutionary algorithms," Z. Michalewicz, M. Schoenauer, pp. 264–269 in *Encyclopedia of Operations Research and Management Science*, 2nd edition, S. I. Gass, C. M. Harris, editors, Kluwer Academic Publishers, Boston, 2001; *Foundations of Genetic Programming*, R. Poli, W. B. Langdon, Springer-Verlag, New York, 2002]

1975 John von Neumann Prize for fundamental theoretical contributions established

The John von Neumann Prize is given in recognition of the scholar who has made fundamental theoretical contributions to operations research and management science. The first such award (1975) went to George B. Dantzig for his development of linear programming and the simplex method. The award was established jointly by ORSA and TIMS, but is now presented by INFORMS.

1975 The Goodeve Medal

This award, established by the Operational Research Society in honor of Sir Charles Goodeve, is given in recognition of the most outstanding contribution to the philosophy, theory, or practice of OR published in the *Journal* of the Society. It was first awarded in 1976 to B. H. Mahon and R. J. M. Bailey for their paper "A proposed improvement replacement policy for Army Vehicles," *Operational Research Quarterly*, 26, 3i, 1975, 477–494."

1975 ***Queueing Systems, Volume I: Theory***, Leonard Kleinrock, John Wiley & Sons, New York

An outgrowth of the author's course taught over five years at UCLA, this text presents the theory of queues at the first-year graduate level. The first half of the book introduces elementary queueing theory, while the second half covers the $M/G/1$ and $G/M/m$ queueing systems, followed by advanced material on the $G/G/1$ queue. Kleinrock states his desire to strike a balance between theory and application. To avoid the "dull theorem-proof format," he prefers to lead the reader through a series of steps that helps the reader to "discover" the result.

1975 ***Analysis for Public Decisions***, Edward S. Quade, American Elsevier

This book describes a new approach for public decision making that rest heavily on analytic methods. Most of the material stems from researchers at the RAND Corporation.

1975 ***Theory of Optimal Search***, Lawrence D. Stone, Academic Press, New York

Search theory was one of the earliest and important applications of OR to military problems. This book extends and updates the state of the art with respect to the problem of optimal allocation of effort to detect a target, and became the standard treatment of the subject. The ideas in this book have found applications in the search for objects by the U.S. Coast Guard, including sunken submarines and unexploded ordnance. Stone was awarded the 1975 Lanchester prize for this book.

1976 The Association of European Operational Research Societies (EURO) was founded

EURO is the Association of European Operational Research Societies within IFORS. The members of EURO comprise the national OR societies of countries located within or nearby Europe.

1976 Robust quality

Genichi Taguchi's notion of quality involves conformance to optimal or ideal parameters that measure quality characteristics. His key idea is that any deviation from these optimal targets causes a total loss to society that involves not only the manufacturer, but the total chain of affected parties who come into contact with the product. In practice, Taguchi favors a quadratic loss function to show the nonlinear increase in the losses with increases in the deviation from the target values or the magnitude of the variance. Taguchi also stresses the importance of robust product and process design with respect to changes in the environment of production or use. In the 1970s, Taguchi's approach became widespread in Japan. His ideas came into use in the U.S. in 1980s, but not without some controversy. ["Robust quality," G. Taguchi, D. Clausing, *Harvard Business Review*, 90, 1, 1980, 65–75; *Introduction to Quality Engineering*, G. Taguchi, Asian Productivity Organization, American Supplier Institute, Dearborn, 1983; "Scientific quality management and management science," P. J. Kolesar, pp. 671–709 in *Handbooks in Operations Research and Management Science, Vol. 4: Logistics of Production and Inventory*, S. C. Graves, A. H. G. Rinnooy Kan, P. H. Zipkin, editors, North-Holland, Amsterdam, 1983; *Designing for Quality: An Introduction to the Best of Taguchi and Western Methods of Statistical Experimental Design*, R. H. Lochner, J. E. Matar, Quality Resources, New York, 1990]

Genichi Taguchi

1976 *Queueing Systems, Volume II: Computer Applications*, Leonard Kleinrock, John Wiley & Sons, New York

This is the companion volume to the author's 1975 textbook on the theory of queueing systems. Here, the concentration is on such applications as priority queues, computer-communication networks, time-shared computer systems, and packet-switched networks. Much of the material is based on the author's research for the U.S. Department of Defense Advanced Research Projects Agency that led to the ARPANET computer-communication network and the Internet. Leonard Kleinrock is considered to be the "father of the Internet." Kleinrock was awarded the 1976 Lanchester prize for Volume II.

1976 *Combinatorial Optimization: Networks and Matroids*, Eugene L. Lawler, Holt, Rinehart and Winston, New York (Dover reprint 2000)

This text was the first exposition of the exciting developments in combinatorial optimization of the 1960s and 1970s, covering such topics as network flows, matching (bipartite and non-bipartite), matroids, and the matroid intersection algorithm. As the author states in the preface, "The last half of the book exists only because of the pioneering insights of Jack Edmonds." Lawler started to write the book in fall 1968; it took eight years to complete.

1976 ***Models in the Public Policy***, Martin Greenberger, Matthew A. Crenson, Brian L. Crissey, Russell Sage Foundation, New York

The subtitle of this book is "Public Decision Making in the Computer Era." The authors present a historical and critical review of the variety of methods used in the study of public policy issues, and discuss early and important activities such as the New York City-RAND Institute, world models, and econometric modeling.

1976 ***Decisions with Multiple Objectives: Preferences and Value Trade-offs***, Ralph Keeney, Howard Raiffa, John Wiley & Sons, New York

Howard Raiffa Ralph Keeney

"The theory of decision analysis is designed to help an individual make a choice among a set of *prespecified* alternatives." This book was instrumental in establishing the field of decision analysis as a "... prescriptive approach designed for normally intelligent people who want to think hard and systematically about some important real problems." Keeney and Raiffa were awarded the 1976 Lanchester prize for this book.

1977 Anti-cycling rules for linear-programming problem

When solving a linear-programming problem by the simplex method, the iterative process may not converge to a finite optimal solution (given that one exists), with the reason being that a set of basic solution changes were found that kept repeating, that is, the process cycled. One way of avoiding such cycles is to ensure that all simplex-based solutions are strictly positive, the nondegeneracy assumption. To be effective, simplex-based software incorporate nondegeneracy procedures that avoid cycles. In contrast to such nondegeneracy procedures, Robert Bland developed simplex method anti-cycling rules that also guarantee that no cycles can occur. Although these rules are of theoretical interest and easy to implement, they are rarely incorporated into simplex method software, as doing so tends to increase the solution time. ["New finite pivot rules for the simplex method," R. G. Bland, *Mathematics of Operations Research*, 2, 2, 1977, 103–107]

1977 The EURO *European Journal of Operational Research* (EJOR) begins publication

This journal is the joint publication of those countries that form the EURO consortium of OR societies that belong to IFORS. The first editors, an editorial triumvirate, were Alan Mercer, Hans-Jürgen Zimmerman, and Bernhard Tilanus.

1977 25th anniversary of ORSA

The silver anniversary of the founding of ORSA was celebrated at the TIMS/ORSA May meeting in San Francisco.

1977 *Exploratory Data Analysis*, John W. Tukey, Addison-Wesley, Reading

Exploratory data analysis (EDA) is part of descriptive statistics: it provides tools for discovering and summarizing relations between variables. In contrast, the traditional inferential methods of statistics are confirmatory in nature: they are well suited to testing and confirming hypotheses once the relations have been formulated. Led by John W. Tukey of Princeton University, the EDA approach gained recognition in the early 1970s. This textbook, written for first-year students at Princeton, assumes no prior knowledge of statistics; it contains a wealth of novel approaches to the exploration of data.

1977 *Manpower Planning Models*, Richard C. Grinold, Kneale T. Marshall, North-Holland, New York

Manpower planning is a temporal model that tries to ensure that the right number of people are available with the right skills at the right time. Manpower models typically track the flows of personnel categories within a system over time. The dynamics of the system is governed by rates of transitions among skill classes or ranks, as well as recruitment and departures. Historically, manpower planning has been used widely in the military sector. The authors introduce the essential elements of manpower planning and cover many applications of the 1960s and 1970s and related issues (data and model validation), much of it based on their research. The history of the subject in the United Kingdom is described by Smith and Bartholomew (1988). ["Military manpower planning models," S. I. Gass, *Computers & Operations Research*, 18, 1, 1991, 65–73; "Manpower planning in the United Kingdom: An historical review," A. R. Smith, D. J. Bartholemew, *Journal of the Operational Research Society*, 9, 1988, 235–248]

1977 *Models for Public Systems Analysis*, Edward J. Beltrami, Academic Press, New York

This text focuses on the delivery of urban services with a special emphasis on model formulation and implementation. A strong feature of the book is the applications it reviews. In particular, discussions include the urban service models of Richard Larson, and the joint work of the author and Lawrence D. Bodin for New York City's sanitation department, including their study of municipal waste collection (Beltrami and Bodin, 1974). ["Networks and vehicle routing for municipal waste collection," E. Beltrami, L. Bodin, *Networks*, 4, 1974, 65–94]

1978 Data envelopment analysis (DEA)

Data envelopment analysis is a procedure for evaluating the relative performance (efficiency) of a set of entities [decision making units (DMUs) such as hospitals, banks, schools] that are responsible for converting inputs (personnel, funds) into outputs (patients processed, customers served, graduates). The process is based on a linear-programming primal–dual structure in which each DMU can be compared to all others in terms of inputs and outputs so as to determine the relative efficiency of each DMU. Although originally developed for nonprofit organizations, DEA has been successfully used to evaluate profit-making DMUs. ["Measuring efficiency of decision making units," A. Charnes, W. W. Cooper, E. Rhodes, *European Journal of Operational Research*, 2, 6, 1978, 429–444; "Data envelopment analysis," W. W. Cooper, pp. 183–191 in *Encyclopedia of Operations Research and Management Science*, 2nd edition, S. I. Gass, C. M. Harris, editors, Kluwer Academic Publishers, Boston, 2001]

8

Methods, applications, technology, and publications from 1979 to 2004

1979 Prospect theory

Daniel Kahneman and Amos Tversky started their investigations of the psychology of human judgment in the early 1970s, focusing particularly on deviations from rationality and judgment heuristics. These heuristics, as identified by Kahneman and Tversky, are "rules of thumb" that people use to tackle a difficult judgmental problem when they lack the cognitive mechanisms to readily solve the problem with precision. A result of the long-time Kahneman and Tversky collaboration is prospect theory, a theory they devised to account for deviations of decision makers from the standard normative expected utility theory. Prospect theory holds that people typically do not monitor the impact of a prospect on their final asset position or total wealth. Rather, they evaluate the outcome of a course of action in terms of the gain or loss relative to a reference point. Moreover, they are highly sensitive to how choices are presented or "framed." Prospect theory also recognizes the asymmetry between gains and losses: The pain generated by a loss tends to exceed the amount of pleasure produced by an equally large gain. The Kahneman–Tversky research program on rationality has shaped current scholarship and research in medicine, law, public policy, international relations, decision analysis, and economics. In recognition of his joint work with Tversky, Kahneman was awarded the 2002 Nobel prize in economics (joint with Vernon L. Smith) for having integrated insights from psychological research into economic science, especially concerning human judgment and decision making under uncertainty. Tversky died in 1996 and was thus not eligible for the Nobel. ["Judgment under uncertainty: Heuristics and biases," A. Tversky, D. Kahneman, *Science*, 185, 1974, 1124–1131; "Prospect theory: An analysis of decision under risk," D. Kahneman, A. Tversky, *Econometrica*, 47, 2, 1979, 263–291; "The framing of decisions and the psychology of choice," A. Tversky, D. Kahneman, *Science*, 211, 1981, 453–458; "Choice theory," *Judgment under Uncertainty: Heuristics and Biases*, D. Kahneman, P. Slovic, A. Tversky, editors, Cambridge University Press, Cambridge, U.K., 1982; "Advances in prospect theory: Cumulative representation of uncertainty," A. Tversky, D. Kahneman, *Journal of Risk Uncertainty*, 5, 1992, 297–323; "Tversky, Amos," T. Gilovich, pp. 849–850 in *The MIT Encyclopedia of the Cognitive Sciences*, R. A. Wilson, F. Keil, editors, MIT Press, Cambridge, 1999]

Amos Tversky
1937–1996

(©Nobel Foundation)
Daniel Kahneman

1979 Spreadsheets and OR add-in software

The first personal computer spreadsheet, VisiCalc, was introduced in October, 1979. It was first conceived by Daniel Bricklin, with the help of Robert Frankston, and marketed by Personal Software, Inc., directed by Daniel Fylstra. However, subsequent use and importance of spreadsheets resulted from Lotus Development Corporation's Lotus 1-2-3 spreadsheet system that combined worksheets with graphics and data-base capabilities. Since that time, OR and related analytical techniques have been included in the major spreadsheet programs such as Excel, Lotus 1-2-3 and Quattro Pro. These add-ins enable the user to do statistical analyses, optimization (linear, integer and nonlinear programming), simulation, decision analysis, forecasting, and financial engineering. *What's Best*, developed by Sam L. Savage (1984), Kevin Cunnigham, and G. Link, was first to combine spreadsheet technology with a linear-programming solver to form an integrated software package, with later enhancements enabling other OR-related analyses to be made. Spreadsheets have become an important computational tool for business managers and analysts, as well as a pedagogical platform. [*What's Best: Takes Your Spreadsheet Beyond "What If"*, Sam L. Savage, LINDO Systems Inc., Chicago, 1984; *Management Science: A Spreadsheet Approach*, D. R. Plane, Boyd & Fraser, Danvers, 1994; "Spreadsheets," D. R. Plane, pp. 780–782 in *Encyclopedia of Operations Research and Management Science*, 2nd edition, S. I. Gass, C. M. Harris, editors, Kluwer Academic Publishers, Boston, 2001]

1979 *Computers and Intractability: A Guide to the Theory of NP-Completeness*, Michael R. Garey, David S. Johnson, W. H. Freeman and Co., New York

An entire generation of OR researchers learned the fundamentals of complexity theory and its implications for optimization algorithms from this classic book. Of particular interest was the compendium of problems in which the authors summarized the known complexity results for an impressive list of optimization and recognition problems. An extended list of this kind appears in Ausiello et al. (1999). Garey and Johnson were awarded the 1979 Lanchester prize for this book. [*Complexity and Approximation: Combinatorial Approximation Problems and Their Approximability Properties*, G. Ausiello, P. Crescenzi, G. Gambosi, V. Kann, A. Marchetti-Spaccamela, M. Potasi, Springer-Verlag, Berlin, 1999]

1980 Flexible manufacturing systems

A flexible manufacturing system (FMS) consists of several computer-controlled machine tools, each capable of performing many operations, that are linked with automated material handling equipment. OR techniques (queueing networks; linear, integer, and nonlinear programming; simulation; heuristic algorithms) have been used to resolve FMS planning problems (set-up decisions) and FMS scheduling problems (real-time scheduling of the manufactured parts). FMSs were operational in the early 1970s, especially in the metal-workings industry, but they did not become a field of study by OR analysts until the late 1970s, with the first research papers appearing in the 1980s. ["Formulation and solution of nonlinear integer production planning problems for flexible manufacturing systems," K. E. Stecke, *Management Science*, 29, 3, 1983, 273–288; "Design, planning, scheduling, and control problems of flexible manufacturing systems," K. E. Stecke, *Annals of Operations Research*, 3, 1985, 3–12; "Flexible manufacturing systems," K. E. Stecke, pp. 226–229 in *Operations Research and Management Science*, 1st edition, S. I. Gass, C. M. Harris, editors, Kluwer Academic Publishers, Boston, 1996]

Kathryn E. Stecke

1980 Constraint programming

Constraint programming (constraint logic programming) originated in computer science and artificial intelligence. Constraint programming techniques have been shown to be effective for solving optimization problems, especially those that arise in sequencing and scheduling, and, in general, combinatorial-structured problems (integer programming). Constraint programming formulates the problem within a programming language and, in the search for an optimal solution, uses logic-based methods to reduce the solution space. ["Constraint programming," I. Lustig, J.-F. Puget, pp. 136–141 in *Encyclopedia of Operations Research and Management Science*, 2nd edition, S. I. Gass, C. M. Harris, editors, Kluwer Academic Publishers, Boston, 2001; "Logic, optimization, and constraint programming," J. N. Hooker, *INFORMS Journal of Computing*, 14, 4, 2002, 295–321]

1980 The Analytic Hierarchy Process

Procedures for resolving multicriteria decision problems all require some process by which the criteria are assigned weights, with the alternatives then compared against the criteria so as to proportionally allocate the criteria weights to the alternatives. Such procedures are often based on what seem to be, at least to the developers, "reasonable" heuristic and/or mathematical procedures. No one method fits all problems and no one method can be said to be the best. But, the Analytic Hierarchy Process (AHP), developed by Thomas

L. Saaty, has proven to be most appropriate for a very wide class of applications. The AHP is based on a sequence of pairwise comparisons of the criteria that leads to the determination of a set of criteria weights, and another sequence of pairwise comparions of the alternatives with respect to the criteria to determine associated weights for the alternatives. The comparisons are made using a fundamental numerical (ratio) scale; the final weights that rank the alternatives are also ratio scale numbers. Thus, the rankings can be compared in a numerically correct fashion, that is, we can determine how much better is one alternative with respect to another. The weights, based on the numerical comparisons, are mathematically derived by computing the associated eigenvector. It can be shown that if the comparisons are consistent, that is, satisfy fundamental transitivity conditions, then the resulting weights are the true weights. If transitivity is not maintained, then the associated eigenvector approximates the unknown true weights, given that readily computed inconsistency information is not extreme. The AHP has been extended to the Analytic Network Process (ANP) which enables more complex multcriteria problems to be resolved. [*The Analytic Hierarchy Process*, T. L. Saaty, McGraw-Hill, New York, 1980; *The Analytic Network Process*, T. L. Saaty, RWS Publications, Pittsburgh, 1996; "The analytic hierarchy process," and "The analytic network process," T. L. Saaty, pp. 19–35 in *Encyclopedia of Operations Research and Management Science*, 2nd edition, S. I. Gass, C. M. Harris, editors, Kluwer Academic Publishers, Boston, 2001; *Decision by Objectives*, E. H. Forman, M. A. Selly, World Scientific Publishing Company, River Edge, 2001; "On teaching the analytic hierarchy process," S. I. Gass, L. Bodin, *Computers & Operations Research*, 30, 10, 2003, 1487–1497]

1980 Decomposition theorem for totally unimodular matrices

Paul D. Seymour's characterization of totally unimodular (TU) matrices is a deep and beautiful result of combinatorial analysis. It shows that any TU matrix arises from network matrices and a couple of special matrices through certain matrix compositions. It follows that there is a good characterization for the problem: "Is a given matrix TU?" This leads to a polynomial algorithm for testing for total unimodularity. ["Decomposition of regular matroids," P. D. Seymour, *Journal of Combinatorial Theory*, B, 28, 1980, 305–359; see also Chapters 19 and 20 of *Theory of Linear and Integer Programming*, Alexander Schrijver, John Wiley & Sons, New York, 1986]

1980 LINDO (Linear and Discrete Optimization) software

Conceived and developed by Linus E. Schrage for mainframe computers, LINDO software for solving linear and integer programming problems had a strong influence in the application and future development of optimization software. PC LINDO, developed by Kevin Cunningham, became available in 1982, with subsequent successful use, especially in the classroom. [*Optimization Modeling with LINDO*, L. Shrage, Duxbury Press, Pacific Grove, 1997]

Linus E. Schrage

1980 Yield (revenue) management

Perishable products such as airline seats are worthless if not utilized by flight time. The idea behind yield management is, based on past and forecasted demand data, to dynamically change the prices for perishable products so as to maximize revenues. Implemented by American Airlines in the 1980s, it has proven to be an effective process that combines OR and artificial intelligence procedures. The process has since been used by other purveyors of perishable products such as hotels, cruise lines, and passenger railroads. ["SABRE soars," T. M. Cook, *OR/MS Today*, 3, 1998, 26–31; "Yield management," R. O. Mason, S. A. Conger, p. 391 in *Encyclopedia of Operations Research and Management Science*, 2nd edition, S. I. Gass, C. M. Harris, editors, Kluwer Academic Publishers, Boston, 2001]

1981 The personal computer

Although prototype and commercially available personal computers were available prior to 1981 (Atari, Apple), it was IBM's unveiling of its personal computer on August 12, 1981 that changed the world-view of how computers can be of service to the general population. From an OR perspective, it brought the means of hands-on analysis to the OR professional, academic and student. The proliferation of OR-related, PC-based software has had the positive benefit of ready availability by the informed user, along with the negative concern of ready analysis by the untrained novice. The use of PCs with its imbedded spreadsheets and analytical add-ons has improved markedly the ability of the OR analyst to conduct and carry out both research and applied projects. [*Big Blues: The Unmaking of IBM*, P. Carroll, Crown Publishers, New York, 1993; *Fire in the Valley: The Making of the Personal Computer*, P. Freibergh, M. Swaine, McGraw-Hill, New York, 1999]

1981 *Urban Operations Research*, Richard C. Larson, Amadeo Odoni, Prentice-Hall, Englewood Cliffs

This book grew out of the course "Analysis of Urban Systems" which the authors started to teach at MIT in 1971. The purpose of the course and the text was to bring the OR methodological toolkit to a set of applications selected from urban systems and services. Unique in its coverage, the text is especially strong in exploring the relations between geometry and probability. It covers spatially distributed queues and location and routing problems. Also of interest is a final chapter on implementation, complete with mini-war stories from the trenches. ["Public sector Operations Research: A personal journey," R. C. Larson, *Operations Research*, 50, 1, 2002, 135–145]

1981 Simulated annealing

Simulated annealing is an optimization method motivated by principles of statistical physics. Physical systems can be steered towards a global minimum energy state by an annealing process, whereby the temperature is slowly lowered, thus allowing the system to attain a metastable equilibrium at each temperature. The application of this principle to combinatorial optimization problems, where the minimum state corresponds to the minimum value of an objective function, was suggested by S. Kirkpatrick, C. D. Gelatt, M. P. Vecchi. ["Optimization by simulated annealing," S. Kirkpatrick, C. D. Gelatt, M. P. Vecchi, *Science*, 220, 1981, 671–680; "Simulated annealing," G. Anandalingam, pp. 748–751 in *Encyclopedia of Operations Research and Management Science*, 2nd edition, S. I. Gass, C. M. Harris, editors, Kluwer Academic Publishers, Boston, 2001]

1981 Equivalence of separation and optimization

The separation problem for a given polyhedron P in $\mathbf{R}^n$ defined by rational linear inequalities is as follows: Given a vector y with rational components, decide whether y belongs to P. If not, find a vector d such that $dx < dy$ for all x in P. Martin Grötschel, László Lovász, and Alexander Schrijver proved that if the separation problem can be solved in polynomial time, so can the optimization problem $\text{Min}\, cx$ over P. This result, which is a theoretical consequence of the ellipsoid method for linear programming, leads to new insights and results for certain combinatorial problems. Grötschel, Lovász, and Schrijver were awarded the 1982 Fulkerson prize for their 1981 paper on this subject. The Fulkerson Prize, awarded for outstanding papers in the area of discrete mathematics, is sponsored jointly by the Mathematical Programming Society (MPS) and the American Mathematical Society (AMS). ["The ellipsoid method and its consequences in combinatorial optimization," M. Grötschel, L. Lovász, A. Schrijver, *Combinatorica*, 1, 1981, 169–197; *Geometrical Algorithms and Combinatorial Optimization*, M. Grötschel, L. Lovász, A. Schrijver, Springer-Verlag, Berlin, 1988]

1981 *Matrix-Geometric Solutions in Stochastic Models: An Algorithmic Approach*, Marcel F. Neuts, The Johns Hopkins University Press, Baltimore (Dover reprint 1994)

An outgrowth of lectures given in July 1979 at John Hopkins University, this book presents a unified computational approach for a variety of queueing and stochastic problems. Distributions of the phase type (PH), which are generalizations of the Erlang family of distributions, are introduced in chapter 2 and play an important role in the $GI/PH/1$ queueing model analyzed in detail in chapter 4. An underlying theme of the book, also developed in Neuts (1986), is that in addressing probability models, researchers need to attach a higher value to the examination of the algorithmic aspects of the solution technique proposed, and that often a satisfactory solution is only obtained if the importance of the computational issues is duly recognized. ["An algorithmic probabilist's apology," M. F. Neuts, pp. 213–221 in *The Craft of Probabilistic Modeling: A Collection of Personal Accounts*, J. Gani, editor, Springer-Verlag, New York, 1986]

Marcel F. Neuts

1982 Algebraic modeling languages

The high-speed of computers, advances in memory, and the refinements in mathematical-programming systems (solvers) enable us to solve real-world mathematical-programming problems with thousands of constraints and many thousands of variables. The question arises: How do we generate such problems so they are accepted by the computer software and are in a form that enables one to show that the problem (model) is correct? It is an imposing task to combine the data and the problem constraints into an explicit form and then enter them into the computer. From the analyst's perspective, the necessary problem statement is best given in an algebraic manner. But solvers require an explicit problem statement. Algebraic modeling languages close the gap between the modeler and the computer by taking an algebraic, concise statement of the problem and data, and generating the requisite format for the solver. The first such language was the General Algebraic Modeling System (GAMS) developed by Johannes Bisschop and Alexander Meeraus. Other algebraic languages include AIMMS (J. Bisschop, R. Entrike); LINGO (K. Cunningham, L. Schrage); AMPL (R. Fourer, D. M. Gay, B. W. Kernighan); MathPro (D. Hirshfeld); MPL (B. Kristjansson). The algebraic language enables the analyst to grow the model in steps, from small size to large, thus facilitating the verification that the final statement of the problem is correct. ["On the development of a general algebraic modeling system in a strategic planning environment," J. Bisschop, A. Meeraus, *Mathematical Programming Study*, 20, 1982, 1–29; "Algebraic modeling languages for optimization," R. Rosenthal, pp. 16–19 in *Encyclopedia of Operations Research and Management Science*, 2nd edition, S. I. Gass, C. M. Harris, editors, Kluwer Academic Publishers, Boston, 2001]

Johannes Bisschop

1982 Average running time of the simplex method

Although the simplex method has been shown to be an exponential-time algorithm, its use in practice indicates that it is very efficient. As an indicator of this behavior, Karl-Heinz Borgwardt showed that on the average the simplex method is a polynomial-time algorithm. ["Some distribution-independent results about the asymptotic order of the average number of pivot steps of the simplex method," K.-H. Borgwardt, *Mathematics of Operations Research*, 7, 1982, 441–462; "The average number of pivot steps required by the simplex method is polynomial," K.-H. Borgwardt, *Zeitschrift für Operations Research*, 26, 1982, 157–177; "On the average number of steps of the simplex method of linear programming," S. Smale, *Mathematical Programming*, 27, 3, 1983, 241–262; *The Simplex Method – A Probabilistic Approach*, K.-H. Borgwardt, Springer-Verlag, Berlin, 1987]

1982 The Roundtable founded

To better serve OR/MS practitioners and its institutional members, TIMS established the Roundtable. A main purpose of the Roundtable is to provide a proactive forum in which leading practitioners discuss matters of mutual interest and undertake cooperative efforts.

Its membership is comprised of institutions, not individuals. The Roundtable is now a part of INFORMS. [http://roundtable.informs.org/Founding.html]

1982 *Fair Representation: Meeting the Ideal of One Man, One Vote*, Michel L. Balinski, H. Peyton Young, Yale University Press, New Haven (second edition, Brookings Institute, 2001)

The authors address the apportionment problem of fairly dividing the seats in a legislature according to the populations of federal states or party votes. The apportionment issue created controversy as far back as 1791, when Jefferson and Hamilton proposed different procedures for apportionment. (Washington exercised the first presidential veto when he disagreed with the support Congress lent to Hamilton's method.) Combining the history, politics, and mathematics associated with apportionment, Michel L. Balinski and H. Peyton Young develop a theory of fair representation that establishes various principles for translating state populations – or vote totals of parties – into a fair allocation of congressional seats. It turns out that most "reasonable" algorithms for apportionment are biased and can produce paradoxical results. For example, a given state can get less seats if the total number of seats increases for the entire body and the populations of the states remain unchanged. The authors develop an impossibility theorem that shows that there is no perfect method; a compromise must be made.

1982 *The Art and Science of Negotiation*, Howard Raiffa, Harvard University Press, Cambridge

This book stems from the author's MBA course in "Competitive Decision Making" and his experiences as the first director of the International Institute for Applied Systems Analysis. "A sophisticated self-help book," it is written for the general audience and uses many real-world (Camp David, Panama Canal) and other negotiation cases to describe the art and science of dispute resolution. As Raiffa noted (2002), this was an "early attempt to show how analysis can be an integral part of the theory and practice of negotiations." The book by Young (1991) contains additional results in game theory, economics, and psychology, as applied to the negotiation process. [*Negotiation Analysis*, H. P. Young, editor, The University of Michigan Press, Ann Arbor, 1991, "Decision analysis: A personal account of how it got started and evolved," H. Raiffa, *Operations Research*, 50, 1, 2002, 179–185]

1983 Integer programming with fixed number of variables

Using methods from the geometry of numbers, Hendrik W. Lenstra, Jr. devised an efficient algorithm for basis reduction. He then proved that for any fixed n, the integer programming problem Max $\{cx \mid Ax \leqslant b,\ x_i$ integer for $i = 1, \ldots, n\}$ can be solved in polynomial time. ["Integer programming with a fixed number of variables," H. W. Lenstra, Jr., *Mathematics of Operations Research*, 8, 1983, 538–548; *Geometric Algorithms and Combinatorial Optimization*, M. Grötschel, L. Lovász, A. Shrijver, Springer-Verlag, New York, 1988]

1984 Interior point methods

The solution space to a linear-programming problem is a convex polyhedron. The simplex method moves along the edges of the polyhedron, with each step (pivot) finding an extreme point (corner) of the polyhedron. Although this may seem to be an inefficient process due to the possibility of the polyhedron having a large (exponential) number of such points, the simplex method works remarkably well in practice. Early researchers in linear programming recognized that an effective way for finding an optimal solution could be that of moving through the interior of the polyhedron. Although certain techniques were developed earlier, the interior-point algorithm of Narinda Karmarkar was the first such polynomial-time procedure. Karmarkar's algorithm and its variations have since proven to be computationally competitive to the simplex method for some large-scale linear-programming problems. Many linear programming computer-based systems incorporate both the simplex method and interior-point procedures. ["A new polynomial-time algorithm for solving linear programming," N. Karmarkar, *Combinatorica*, 4, 1985, 373–395; *Linear Programming*, G. B. Dantzig, M. N. Thapa, Springer-Verlag, New York, 1997]

1984 Neural networks for optimization (Hopfield network)

Neural networks were originally proposed in an attempt to model a certain architecture of connectivity in the mammalian brain, analogous to the connections among neurons. John J. Hopfield suggested that a network of nodes (corresponding to neuron-like elements) with symmetric random connection (synaptic) weights is similar to magnetic material called spin glass which can store different spin patterns. Hopfield then used Donald O. Hebb's modification of synaptic weights to stabilize net activity and find a stable configuration. This led to the structure known as a Hopfield network. In OR, the design of a neural network algorithm involves the choice of a (typically multi-layer) network architecture and a training or learning process to determine and update the weights on the connections between pairs of nodes in the network to train the neural network to recognize good solutions. To solve optimization problems with a neural network algorithm, the weights are adjusted until a stable state is reached that corresponds to a local minimum of the objective function being considered. Hopfield and David W. Tank (1985) constructed a neural net for the traveling salesman problem with n^2 neurons, where the output of neuron (m, k) represents whether city m should be the kth city visited in the tour. [*The Organization of Behavior*, D. O. Hebb, John Wiley & Sons, New York, 1949; "Neurons with graded response have collective computational properties like those of two-state neurons," J. J. Hopfield, *Proceedings of the National Academy of Sciences*, 81, 1984, 3088–3092; "Neural computation of decisions in optimization problems," J. J. Hopfield, D. W. Tank, *Biological Cybernetics*, 52, 1985, 141–152; "Collective computation in neuron like circuits," D. W. Tank, J. J. Hopfield, *Scientific American*, 257, 6, 1987, 104–114; "Neural nets and artificial intelligence," J. D. Cowan, D. H. Sharp, *Daedalus*, 117, 1, 1988, 85–122; *Neural Networks*, J. A. Freeman, D. M. Skapura, Addison-Wesley, Reading, 1991; "Neural networks and operations research: An overview," L. I. Burke, J. P. Ignizio, *Computers and Operations Research*, 19, 3/4, 1992, 179–189; "Neural networks," R. Wilson, R. Sharda, *OR/MS Today*, 19, 4, 1992, 36–42]

1984 The Ramsey Medal

The highest award of the ORSA Special Group on Decision Analysis, the Ramsey Medal, recognizes distinguished contributions to the field of decision analysis. Howard

Raiffa was the first Ramsey medalist in 1984. Frank P. Ramsey was a philosopher/logician with a broad interest in foundations. He gave the first rigorous proof of the expected utility hypothesis proposed by Daniel Bernoulli in 1732. ["Truth and probability," F. P. Ramsey, pp. 23–52 (reprint) in *Studies in Subjective Probability*, H. E. Kyberg, Jr., H. E. Smokler, editors, John Wiley & Sons, New York, 1964; *The Economics of Uncertainty*, K. H. Borch, Princeton University Press, Princeton, 1968; *Decision Analysis*, H. Raiffa, Addison-Wesley, Reading, 1968; "Frank Plumpton Ramsey," Peter Newman, pp. 186–197 in *The New Palgrave: Utility and Probability*, J. Eatwell, M. Milgate, P. Newman, editors, W. W. Norton & Co., New York, 1990]

Truth and consequences:

According to Raiffa (1968), Ramsey, based on his "Truth and probability" paper (1926), is considered to be the first one "... to express an operational theory of action based on the dual, intertwining notions of judgmental probability and utility."

Frank P. Ramsey
1903–1930

1984 ***What's Best: Takes Your Spreadsheet Beyond "What If,"*** Sam L. Savage, LINDO Systems Inc., Chicago

Sam L. Savage

The first computer-based system that combined the power of spreadsheets with optimization procedures for solving linear and nonlinear programming problems.

1985 Power-of-two solutions for lot sizing

Suppose that that the reorder interval T in an inventory replenishment problem is restricted to be of the form $T = 2^k T_b$, where T_b is some fixed base planning period (say, a week) and k a positive integer. Such an ordering policy is called a power-of-two solution. For the standard single-item EOQ lot sizing model, the optimal solution subject to this restriction is no worse than 6% above the true optimum and the average relative error is 2%. The importance of power-of-two policies is that the good error bounds continue to hold when the inventory problem is generalized to more sophisticated settings, such as a serial production system where the reorder cycle of each stage is restricted to be of the form $2^k T_b$ for some integer k. Robin Roundy's results on power-of-two policies showed that the analysis of heuristics with guaranteed performance bounds can produce useful results for notoriously difficult multi-stage or multi-level distribution problems. Roundy was awarded the 1988 Lanchester prize for his 1986 paper. ["98%-effective integer-ratio lot-sizing for one-warehouse multi-retailer systems," R. O. Roundy, *Management Science*, 31, 11, 1985, 1416–1430; "A 98%-effective lot-sizing rule for a multi-product, multi-stage production/inventory system," R. O. Roundy, *Mathematics of Operations Research*, 11, 4, November 1986, 699–727; "Analysis of multistage production systems," J. A. Muckstadt, R. O. Roundy, pp. 59–131 in *Handbooks in Operations Research & Management Science, Vol. 4: Logistics of Production and Inventory*, S. C. Graves, A. H. G. Rinnooy Kan, P. H. Zipkin, editors, North-Holland, New York, 1993]

1985 Branch-and-cut procedures for combinatorial optimization

For an integer programming problem, let P denote the convex hull of all integer solutions to the problem. Typically, a full description of P is not available. A branch-and-cut method uses constraint generation to obtain a sequence of approximations to P. At each step, the linear programming problem providing the current description of P is solved. The solution x^* to this problem is optimal if it is feasible for the original integer problem. Otherwise, one can find violated constraints that separate x^* and P. These are then added to the linear programming approximation and the problem is re-solved. Branching is used only when such violated constraints cannot be generated easily. The power of branch-and-cut derives from the knowledge of facet-generating constraints for P. As a result, branch-and-cut has been successfully applied to combinatorial optimization problems where the structure of the integer polyhedron is well understood. Martin Grötschel and Manfred Padberg had studied this structure for the traveling salesman problem since the mid-1970s. By the mid-1980s, Padberg and his coworkers demonstrated that branch-and-cut is the most promising method for obtaining exact solutions to large traveling salesman problems. A description of this procedure for a traveling salesman problem with 2392 nodes is given in Padberg (1999). ["Polyhedral computations," M. Padberg, M. Grötschel, pp. 307–360 in *The Traveling Salesman Problem: A Guided Tour of Combinatorial Optimization*, E. L. Lawler, J. K. Lenstra, A. H. G. Rinnooy Kan, D. B. Shmoys, editors, John Wiley & Sons, New York, 1985; "Optimization of a 532-city symmetric traveling salesman problem by branch-and-cut," M. Padberg, G. Rinaldi, *Operations Research Letters*, 6, 1987, 1–7; "A branch-and-cut algorithm for the resolution of large-scale symmetric traveling salesman problems," M. Padberg, G. Rinaldi, *SIAM Review*, 33, 1991, 60–100; *Linear Optimization and Extensions*, M. Padberg, Springer-Verlag, New York, 1999]

1985 *Handbook of Systems Analysis: Overview of Uses, Procedures, Applications, and Practice*, Hugh J. Miser, Edward S. Quade, editors, Wiley, Chichester

Hugh J. Miser
1917–1999

This is the first of a three volume discussion of the state-of-the art of systems analysis as viewed by two pioneers of the field. Miser was president of ORSA in 1962. [*Handbook of Systems Analysis: Craft Issues and Procedural Choices*, H. J. Miser, E. S. Quade, editors, Wiley, Chichester, 1988; *Handbook of Systems Analysis: Cases*, H. J. Miser, editor, Wiley, Chichester, 1994]

1986 Tabu search

Tabu search is a metaheuristic that guides a local search procedure in exploring the solution space by adaptively modifying the search neighborhood as the search progresses. The modifications seek to avoid local optima or undue exploration of unattractive regions of the search space. The name tabu refers to the use of memory structures to exclude certain solutions, or regions of the solution space, from the search neighborhood. The central issue of tabu search is the design of memory structures that reinforce actions that lead to good solutions and discourage the actions that result in poor performance. Fred Glover's 1986 paper was the first to use the term tabu search. ["Future paths for integer programming and links to artificial intelligence," F. Glover, *Computers & Operations Research*, 13, 5, 1986, 533–549; "Tabu search – Part I," F. Glover, *ORSA Journal on Computing*, 1, 3, 1989, 190–206; *Tabu Search*, F. Glover, M. Laguna, Kluwer Academic Publishers, Boston, 1997; "Tabu search," F. Glover, pp. 821–827 in *Encyclopedia of Operations Research and Management Science*, 2nd edition, S. I. Gass, C. M. Harris, editors, Kluwer Academic Publishers, Boston, 2001]

Fred Glover

1986 The Harold Larnder Prize

Robert E. (Gene) D. Woolsey

This prize, established by the Canadian Operational Research Society (CORS), is given each year to an individual who has achieved international distinction in OR. The first recipient was Robert E. (Gene) D. Woolsey. Harold Larnder was a Canadian who worked in Great Britain at the Bawdsey Manor Research Station prior to and during WW II. He is considered a co-developer of radar, and helped to turn it into an effective air defense system during the Battle of Britain. He was President of CORS in 1966–1967.

1986 *Theory of Linear and Integer Programming*, Alexander Schrijver, John Wiley & Sons, New York

This book presents rather detailed and clear expositions of the theoretical aspects of linear and integer programming, and provides the reader with valuable historical notes. In addition to a comprehensive treatment of the standard results, Schrijver gives detailed accounts of the major theoretical advances of the early 1980s, starting with the ellipsoid method and its theoretical implications. Strong features of the book include the interplay between complexity theory and polyhedral optimization, and the elegant unification and derivation of earlier results. Schrijver was awarded the 1986 Lanchester prize for this book.

1988 American Airlines Decision Technologies

Thomas M. Cook

Starting in 1982, under the direction of Thomas M. Cook, American Airlines' OR staff was instrumental in pioneering major OR applications. In 1988, the OR group was established as a separate division, American Airlines Decision Technologies (AADT). Robert Crandall, CEO of AA's parent corporation, AMR, credits a long list of OR-based methods and systems development by AADT as being a key reason for AA's strong position in the industry. The list includes: yield management, trip reallocation and improvement program, arrival slot allocation, and flight scheduling. Cook was president of INFORMS in 2003. ["Sabre Soars," T. M. Cook, *OR/MS Today*, 3, 1991, 26–31; "Eyes on the Prize," P. Horner, *OR/MS Today*, 4, 1991, 34–38]

1988 *An Introduction to Queueing Networks*, Jean Walrand, Prentice-Hall, Englewood Cliffs

This text was among the first devoted entirely to the subject of queueing networks. Its numerous examples illustrate the use of queueing networks to model computer systems, communication networks, and manufacturing operations. Walrand received the 1989 Lanchester prize for this book.

1989 Supply chain management

Supply chain refers to the collection of entities encountered as goods flow from suppliers to ultimate customer locations. Supply chain management refers to the integration of these elements and the flows between them to ensure that the product is available in the right amounts at the right locations "in order to minimize system-wide costs while satisfying service level requirements" (Simchi-Levi, Kaminsky, Simchi-Levi, 2000). The subject has emerged from a confluence of logistics, operations, inventory, and distribution management and hence, it is hard to date its beginnings precisely. A defining moment in the field was the work done by the Strategic Planning and Modeling group at Hewlett-Packard (HP) in collaboration with Hau Lee and associates at Stanford University. Initially, the project focused on inventory reduction for HP's personal computers and deskjet printers. Subsequent phases led to such key supply chain strategies as design for localization (market differentiation) and postponement (delayed product differentiation). ["Managing supply chain inventory: Pitfalls and opportunities," H. L. Lee, C. Billington, *Sloan Management Review*, 33, 3, 1992, 65–73; "Material management in decentralized supply chains," H. L. Lee, C. Billington, *Operations Research*, 41, 5, 1993, 835–837; "The evolution of supply-chain-management models and practice at Hewlett-Packard," H. L. Lee, C. Billington, *Interfaces*, 25, 5, 1995, 42–63; "Effective management of inventory and service through product and process redesign," H. Lee, *Operations Research*, 44, 1, 1996, 151–159; *Designing and Managing the Supply Chain: Concepts, Strategies, and Case Studies*, D. Simchi-Levi, P. Kaminsky, E. Simchi-Levi, Irwin–McGraw-Hill, Boston, 2000]

1989 *ORSA Journal of Computing* begins publication

In recognition of their strong interrelationship, this journal is dedicated to the publication of papers in the intersection of OR and computer science. Its founding and first editor was Harvey J. Greenberg. It is now published by INFORMS.

1989 *Stochastic Modeling and the Theory of Queues*, Ronald W. Wolff, Prentice-Hall, Englewood Cliffs

The first part of this well-received text is devoted to renewal theory and Markov chains, with the latter part covering queueing theory. Throughout the text, the author stresses the role of time-average and regeneration arguments. PASTA (Poisson Arrivals See Time Averages) plays an important role in its discussion of queueing.

1990 OR and Operation Desert Storm

The U.S. Military Airlift Command (MAC), using OR analysts and techniques for planning and scheduling of cargo, crews and flights, moved 155,000 tons of equipment and 164,000 personnel to Saudi Arabia in 75 days. The airlift continued during the "hot" Desert Shield operations. By the end of the Persian Gulf War, using a new airlift scheduling tool, the OR-based Airlift Deployment Analysis System, MAC scheduled over 11,500 missions and moved over 350,000 passengers. During the initial 40-day air campaign of the Persian Gulf War, OR analysts and methodologies aided in the planning and scheduling of over 100,000 sorties. The subsequent ground war used the OR-based Logistics Release Point resupply technique that reduced the resupply time for a brigade by half, and enabled both the supply unit and supported combat unit to be moving during resupply activities. ["Crisis analysis: Operation Desert Shield," R. Roehrkasse, G. C. Hughes, *OR/MS Today*, 6, 1990, 22–27; "OR goes to war," T. F. Schuppe, *OR/MS Today*, 2, 1991, 36–44; "Mission (not) impossible," R. D. Kraemer, M. R. Hillard, *OR/MS Today*, 2, 1991, 44–45; "Desert Storm," R. Staats, *OR/MS Today*, 6, 1991, 42–65]

1990 Financial engineering/markets

OR has a long history with respect to the application of OR techniques to a wide-range of financial problems, for example, early work in portfolio analysis. Over the years, as new methodologies have been developed and computer speed and capacity have grown, many of the broader and more complex financial problems have come under the purview of OR professionals. It is difficult to place this field in the correct position on the timeline, but the chosen year marks a time when OR was certainly recognized as being important to the resolution of problems that stem from the area known as financial engineering/markets. In addition to portfolio analysis, the problems of interest include: pricing derivatives, trading tactics, funding decisions, strategic problems, regulatory and legal problems. Here, mathematical programming and Monte Carlo simulation techniques are the principle OR tools, with game theory, network analysis, decision trees, inventory control, and Markov chains also finding application. ["Banking," S. A. Zenios, pp. 45–49 in *Encyclopedia of Operations Research and Management Science*, 2nd edition, S. I. Gass, C. M. Harris, editors, Kluwer Academic Publishers, Boston, 2001; "Financial markets," J. Board, C. Sutcliffe, W. Ziemba, pp. 292–299 in *Encyclopedia of Operations Research and Management Science*, 2nd edition, S. I. Gass, C. M. Harris, editors, Kluwer Academic Publishers, Boston, 2001; "Managing risk, reaping rewards," S. A. Zenios, *OR/MS Today*, 6, 2001, 32–36]

1990 *Operations Analysis in the U.S. Army Eighth Air Force in World War II*, Charles W. McArthur, History of Mathematics, Vol. 4, American Mathematical Society, Providence

During World War II, the England-based U.S. Eighth Air Force formed a 47 analyst Operational Research Section that included 18 mathematicians. Since the primary mission of the Eighth Air Force was strategic daylight bombing, the analysts concentrated on evaluating bombing tactics and their results. (The British Bomber Command had a similarly named evaluation unit.) Using as a measure of effectiveness "How many bombs were within 1000 feet of the assigned aiming point," the work of the OR section helped to increase the percentage from less than 15 percent to better than 60 percent. The mathematicians included James W. Alexander, Edwin Hewitt, Ralph D. James, George W. Mackey, and Angus E. Taylor. The author was a bombardier in the Eighth Air Force. ["Operations analysis in the United States Air Force," L. A. Brothers, *Operations Research*, 2, 1, 1954, 1–16]

1991 First ORSA Prize awarded

The ORSA Prize, now the INFORMS Prize, is awarded to companies that have effectively integrated OR into its organizational decision-making processes. The first winners were American Airlines and Federal Express. ["Eyes on the prize," P. Horner, *OR/MS Today*, 4, 1991, 34–38]

1993 *Network Flows: Theory, Algorithms, and Applications*, Ravindra K. Ahuja, Thomas L. Magnanti, James B. Orlin, Prentice-Hall, Englewood Cliffs, NJ

Running over 860 pages in length, this volume is one of the most comprehensive text on network flows. Its goal, according to the authors, is "to bring together the old and the new, and provide an integrative view of theory, algorithms, and applications." The original stimulus for the book was provided by the research program on designing faster network flow algorithms started by Ahuja and Orlin in 1986. The authors were awarded the 1993 Lanchester prize for this book. Magnanti was president of ORSA in 1988 and president of INFORMS in 1999.

Ravindra K. Ahuja

Thomas L. Magnanti

James B. Orlin

1994 Network-Enabled Optimization System (NEOS)

This Internet/web-based system was initiated by Argonne National Laboratory and Northwestern University with the aim of connecting users of optimization technology and providing them with problem-formulating information and software. NEOS is organized into three parts: (1) NEOS Tools – A library of freely available optimization software written by researchers in the NEOS project; (2) NEOS Guide – A collection of information and educational material about optimization, including a guide to optimization software, descriptions of algorithms, application case studies, FAQs for linear and nonlinear programming, and a collection of test problems and technical reports; (3) NEOS Server – A facility for solving optimization problems remotely over the Internet. ["Optimization on the Internet," J. Czyzk, J. H. Owen, S. J. Wright, *OR/MS Today* 3, 1997, 48–51; http://www.mcs.anl.gov/otc/]

1994 ***Markov Decision Processes: Discrete Stochastic Dynamic Programming***, Martin L. Puterman, John Wiley & Sons

This book provides a unified treatment of nearly four decades of theory and applications of Markov decision processes with discrete state space. For his skillful integration of the field's diverse literature into a text and standard reference work, Puterman received the 1995 Lanchester prize.

1995 INFORMS formed by merger of ORSA and TIMS

Since 1974, the two major U.S. professional societies that evolved from the pre-World War II efforts in OR, the Operations Research Society of America (ORSA) and The Institute of Management Sciences (TIMS), jointly sponsored some activities such as national meetings and the publication of some journals. Twenty years later, it was felt by some of the officers and members of both societies that the profession would prosper and societal activities would be made more cost-efficient if the two societies combined into one. Management of the joint activities, especially budgets, and the need for approval by both societies and the joint board, raised operational concerns and was considered to be cumbersome. Further, it was hypothesized that the new society would attract other OR-related societies to join it to form a society that would encompass a broader view of the profession and attract a wider membership. Some members of ORSA and TIMS opposed the merger. They felt that a merged society would be detrimental to the professional base of the individual societies, and the dynamic forces behind each society would be dissipated; the traditions, scope, and pride of membership associated with ORSA and TIMS would be weakened and/or lost by a merger. The opposing members felt that joint management problems could be alleviated by a new look at the joint board structure that would lead to a simplification and an improvement of the management and budgeting of the joint activities. After a long debate, covering a few national meetings, the merger was put to a vote. The merger was approved. The name of the combined societies: the Institute of Operations Research and the Management Sciences (INFORMS). John D. C. Little was the first president of INFORMS. [*Special Report*: "More than a merger," P. Horner; "The case for a change," R. C. Larson, G. Lilien; "An organization for OR/MS and the information & decision sciences in the 21st century, OR/MS Strategic Planning Committee," *OR/MS Today*, 20, 3, October 1993, 34–42; "Not this merger proposal," S. I. Gass, *OR/MS Today*, Feb. 1994, 44–46]

1995 INFORMS Online (IOL)

INFORMS Online was initially established on the web by Jim Bean and Mohan Sodhi as a means of transmitting and gathering information on INFORMS and OR to and from its members. Its first editor was Michael Trick. ["We've come so Far . . . ," M. A. Trick, *OR/MS Today*, 5, 8, 2000; http://www.informs.org]

1995 *Scheduling: Theory, Algorithms, and Systems*, Michael Pinedo, Prentice-Hall, Englewood Cliffs

This text provides a balanced exposition of scheduling theory and systems, reflecting the resurgence of the area in the 1980s. The book is divided into three parts: deterministic scheduling, stochastic scheduling models (an area to which the author has made considerable contributions), and the practical use of scheduling and scheduling systems.

1996 *Encyclopedia of Operations Research and Management Science*, S. I. Gass, C. M. Harris, editors, Kluwer Academic Publishers, Boston (2nd edition, 2001)

The first encyclopedic overview of operations research and management science.

1996 *Monte Carlo: Concepts, Algorithms, and Applications*, George S. Fishman, Springer-Verlag, New York

This sophisticated treatment of the Monte Carlo method is meant to provide a single-volume text on the subject for both graduate students and professional analysts. An underlying theme of the book is the use of effective sampling plans to reduce errors, both statistical and computational. The book prominently features work on Monte Carlo Markov chain sampling, where the sample paths involve dependent observations based on an appropriate Markov chain formulation. Fishman was awarded the 1996 Lanchester prize for this book.

George S. Fishman

1999 IFORS celebrates its 40th anniversary at the 15th triennial conference in Beijing

In contrast to the first international meeting in OR (Oxford, 1957) that had an attendance of 250 delegates from 21 countries, the Beijing meeting attendance was 969 from 53 countries.

1999 *Manufacturing & Service Operations* begins publication

This INFORMS journal is dedicated to publishing articles related to the theory or practice of the production of goods and services, in all of its aspects. Leroy B. Schwarz served as its first editor.

1999 *Smart Choices: A Practical Guide to Making Better Decisions*, John S. Hammond, Ralph L. Keeney, Howard Raiffa, Harvard Business School Press, Cambridge

This book aims to provide a roadmap for effective decision making to the public at large. The authors believe that by stripping away the academic jargon, the benefits of years of research and teaching can be brought to every person. The book emphasizes problem identification and formulation, and includes a chapter on some of the psychological traps that threaten the decision maker.

2000 50th anniversary of the publication of the *Journal of Operational Research*

To celebrate the 50th anniversary of the *Journal of Operations Research*, the editor, John Ranyard, and the editorial board, selected influential papers that appeared in the last 50 years. They were successively issued during the year, starting with the first published article in an OR journal, P. M. S. Blackett's "Operational Research," *Operational Research Quarterly*, 1, 1, 1950, 3–6.

The first among many:

2000 *Foundations of Inventory Management*, Paul H. Zipkin, McGraw-Hill, Boston

This book is arguably the most comprehensive treatment of inventory theory since the 1963 text by Hadley and Whitin. Intended as a text for doctoral students in operations research and related fields, the book emphasizes the coherence of the body of knowledge represented by inventory theory. The topics covered move from simple single-item models to multiple items and multiple locations, to stochastic lead times and demands, through time varying, stochastic demands. By pointing out the various interconnections among different research streams, the book provides an impressively panoramic view of inventory theory.

2001 EURO working group PROMETHEUS on ethics and OR

The purpose of PROMETHEUS is to inspire OR researchers, teachers, students, consultants and decision makers to integrate ethical aspects in all of their OR activities. It was conceived by Jean-Pierre Brans and founded at the EURO XVIII Conference in Rotterdam. Membership is open to the world-wide OR community. Members must pledge to adhere to the Oath of Prometheus, given below. (www.Prometheus.vub.ac.be)

The Oath of Prometheus

As an OR researcher, I request the largest freedom to collaborate with my colleagues and to investigate without any limits all ideas, all techniques and all methods in any field. However, I shall always keep in mind that the results of my research could possibly be used for human purposes.

As an OR teacher, I commit myself:

- to transmit honestly my knowledge and my know-how;
- to respect my colleagues and to collaborate with them in a spirit of dialogue;
- to discuss with my students the consequences of the possible decisions proposed by the OR models.

As a decision-maker, I commit myself to take into account not only my own objectives but also the social, economic and ecological dimensions of the problems.

As a consultant or an analyst, I commit myself to convince the decision-makers to adopt a fair ethical behaviour and to assist them to meet their goals within the limits of sustainable development. I will feel myself free to refuse to provide information or tools, which to my opinion, could bring into danger the social welfare of mankind and the ecological future of Earth.

Jean-Pierre BRANS, Brussels, Belgium, July 2000

Jean-Pierre Brans

2001 50th anniversary of the OR program at the Naval Post Graduate School (NPGS)

Tracing its origins to a school of marine engineering established in 1909 in Annapolis, Maryland, the NPGS was formed and moved to its present campus in Monterey, California in December 1951. Today the student population at the Postgraduate School has grown to 1,800, with students coming from all service branches of the U.S. defense community, as well as from the Coast Guard, the National Oceanic and Atmospheric Administration, and the services of more than 25 allied nations. Over the years, it has been a prime mover in the education and training of military officers in operations analysis and research.

2002 50th anniversary of the founding of ORSA

Marked by the first national meeting of ORSA on November 17, 1952, the INFORMS San Jose meeting, November 17–20, 2002, celebrated 50 years of U.S. OR professional society activities. ORSA merged with TIMS in 1995 to form INFORMS.

2002 50th anniversary of the publication of *Operations Research*

To celebrate the 50th anniversary of the journal *Operations Research*, the editor, Lawrence M. Wein, asked a number of authors to provide personalized reminiscences of their career in OR. Thirty-three such papers appear in Vol. 50, No. 1, 2002. Collectively, these papers provide the reader with a sweeping historical view of many of the major technical and applied advances that have formed OR.

2002 *Stochastic Process Limits: An Introduction to Stochastic-Process Limits and Their Applications to Queues*, Ward Whitt, Springer-Verlag, New York

The study of communication and other queueing systems in operations research is often concerned with the behavior of systems under heavy traffic, where the control of the queueing system presents the greatest challenge. Stochastic-process limits provide simple approximations for complex stochastic processes, particularly in the case of queues under heavy traffic. This book unifies Whitt's over three decades of research. He was awarded the 2003 Lanchester prize for the book.

2003 50th anniversary of the founding of TIMS

Founded on December 1, 1953, TIMS was instrumental in identifying, extending, and unifying scientific knowledge that contributes to the understanding and practice of management. TIMS merged with ORSA in 1995 to form INFORMS.

2003 *Operational Research in War and Peace*, Maurice W. Kirby, World Scientific, London

This book gives an account of Operational Research in Britain from the late 1930s to 1970. It describes OR's beginnings and origins; World War II developments and military applications; post-wartime activities within the Labour Government and the iron, steel, coal mining industries; the diffusion of OR in corporations; OR in the public sector; and the institutional development of OR.

2003 *Combinatorial Optimization: Polyhedra and Efficiency*, Alexander Schrijver, Springer-Verlag, Berlin

Alexander (Lex) Schrijver

This three-volume work provides a comprehensive overview of polyhedral methods and algorithms in combinatorial optimization. The 1,800 pages and its 4,500 references speak to the amazing growth of the field since the pioneering work of Jack Edmonds in the 1960s. The book's eight parts cover: paths and flows; bipartite and nonbipartite matching and covering; matroids and submodular functions; trees and branchings; cliques, stable sets and coloring; multiflows and disjoint paths; and hypergraphs. Each part includes an introductory exposition, an in-depth treatment of the material, and historical notes.

2004 MIT Operations Research Center 50th anniversary

Started in 1952 as the Committee of Operations Research by physics professor and OR pioneer Philip M. Morse, this early university-based OR endeavor evolved into the cross-campus Operations Research Center. The Center sponsors graduate students from many different departments and carries out both theoretical and applied OR studies.

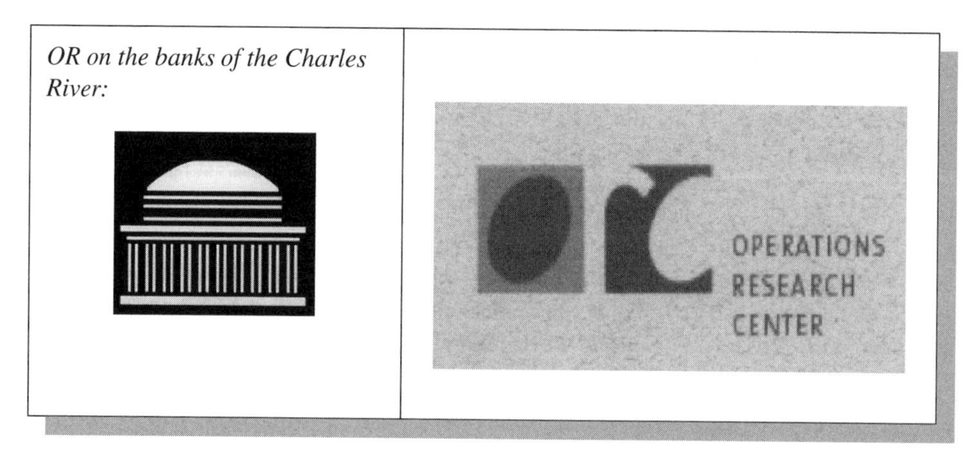

OR on the banks of the Charles River:

Acronyms

ADL Arthur D. Little, Inc.
AGIFORS Airline Group of the International Federation of Operations Research Societies
AHP Analytic Hierarchy Process
AI Artificial Intelligence
AKOR Arbeitskreis Operational Research
ALS Advance Logistics System
AMPL A Modeling Language for Mathematical Programming
AMS American Mathematical Society
ARIMA Autoregressive Integrated Moving Average
ARMA Autoregressive Moving Average
ARPA Advanced Research Projects Agency
ARPANET Advanced Research Projects Network
ASWORG Antisubmarine Warfare Operations Research Group
CEIR Corporation for Economic and Industrial Research
CLT Central Limit Theorem
CNA Center for Naval Analyses
CORS Canadian Operational Research Society
CPC Card Programmed Computer
CPM Critical-Path Method
CPMS College of Practice of Management Science
DEA Data Envelopment Analysis
DMU Decision Making Unit
DSI Decision Sciences Institute
DSS Decision Support System
EDA Exploratory Data Analysis
EJOR *European Journal of Operational Research*
EURO European Operational Research Societies
FMS Flexible Manufacturing Systems
GAMS General Algebraic Modeling System
GOR Gesellschaft fűr Operations Research
GPSS General Purpose Simulations System
IAOR *International Abstracts in Operations Research*
IFORS International Federation of Operational Research Societies
IIASA International Institute for Applied Systems Analysis
INFORMS Institute for Operations Research and the Management Sciences
IOL INFORMS Online
IPL Information Processing Language
JUSE Japanese Union of Scientists and Engineers
KKT Karush–Kuhn–Tucker
LINDO Linear and Discrete Optimization
LISP LISt Processing
LP Linear Programming
LT Logic Theorist

MAC Military Airlift Command
MCDA Multiple Criteria Decision Analysis
MCDM Multiple Criteria Decision Making
MDP Markov Decision Process
MDS Management Decision System
MEDIAC Media Evaluation Using Dynamic and Interactive Applications of Computers
METRIC Multi-Echelon Technique for Recoverable Item Control
MPL Mathematical Programming Language
MPM Metra Potential Method
MPS Mathematical Programming Society
MS Management Science
NEOS Network-Enabled Optimization System
NPGS Naval Post Graduate School
NYCRI New York City RAND Institute
OEG Operations Evaluation Group
OR Operations Research/Operational Research
OR/MS Operations Research/Management Science
ORC Operations Research Center (MIT)
ORG Operations Research Group
ORO Operations Research Office
ORS Operations Research Society (UK)
ORSA Operations Research Society of America
ORSJ Operations Research Society of Japan
PASTA Poisson Arrivals See Time Averages
PERT Program Evaluation and Review Technique
PFI Product Form of the Inverse
PPBS Planning, Programming and Budgeting System
RAC Research Analysis Corporation
ROADEF Association Française de Recherche Opérationelle et d'Aide à la Décision
SAT Satisfiability Problem
SCOOP Scientific Computation of Optimal Programs
SEAC Standards Eastern Automatic Computer
SIAM Society for Industrial and Applied Mathematics
SOFRO Société Française de Recherche Opérationelle
SSM Soft Systems Methodology
SUMT Sequential Unconstrained Minimization Techniques
TIMS The Institute of Management Science
TQM Total Quality Management
TSP Traveling Salesman Problem

Name index

Subject index

SAUL I. GASS

Saul I. Gass received his B.S. in Education and M.A. in Mathematics from Boston University, and his Ph.D. in Engineering Science/Operations Research from the University of California, Berkeley. He is Professor Emeritus at the Robert H. Smith School of Business, University of Maryland, College Park. He is a University of Maryland Distinguished Scholar-Teacher and Dean's Lifetime Achievement Professor for the Robert H. Smith School of Business. Dr. Gass served as a mathematician for the Aberdeen Bombing Mission, U.S. Air Force, and then transferred to Air Force Headquarters where he began his career in operations research with the Directorate of Management Analysis, the organization in which linear programming was first developed. For IBM he was an Applied Science Representative, Manager of the Project Mercury Man-in-Space Program, and Manager of IBM's Federal Civil Programs. Included in his many publications are the texts *Linear Programming* (fifth edition) and *Decision Making, Models and Algorithms*, and the book *An Illustrated Guide to Linear Programming*. He is co-editor of the *Encyclopedia of Operations Research and Management Sciences* and *A Guide to Models in Governmental Planning and Operations*. He is a past president of the Operations Research Society of America (ORSA) and Omega Rho, the international operations research honor society. He is a recipient of ORSA's Kimball Medal for service to the society and the profession, INFORMS's Expository Writing Award, the Military Operations Research Society's Jacinto Steinhardt Memorial Award for outstanding contributions to military operations research. He is a Fellow of INFORMS. He was a 1995–1996 Fulbright Research Scholar.

ARJANG A. ASSAD

Arjang A. Assad received his Ph.D. in Management Science from the Sloan School of Management at MIT in 1978. He also holds a B.S. in Mathematics, and master's degrees in Chemical Engineering and Operations Research from the same institution. He joined the University of Maryland in 1978, where he has been Professor of Management Science since 1988. During 1998–2003, he was chairperson of the Decision & Information Technologies – the largest faculty area within the Robert H. Smith School of Business. Previously, Dr. Assad had directed the IBM Total Quality Project and won the 1996 Maryland Association of Higher Education Award for his role in developing its joint business-engineering curriculum. He received the 2002 Kirwan Undergraduate Education Award, the highest recognition given for contributions to undergraduate education at the University of Maryland. Dr. Assad led the effort to create the Academy of Excellence for Teaching & Learning on his campus and served as its first chair. Dr. Assad is the coauthor of two books: *Excellence in Management Science Practice: A Readings Book*, and *Vehicle Routing: Methods and Studies*. He has also published numerous papers on optimization models for distribution, transportation and manufacturing systems. He has participated in the development of commercial routing software and in a broad spectrum of consulting projects. Dr. Assad has chaired the History and Traditions Committee of INFORMS during 1996–2004. He has also served on the editorial boards of several journals. Dr. Assad was recently named Senior Associate Dean for Academic Affairs at the Smith School of Business.

Early Titles in the
INTERNATIONAL SERIES IN
OPERATIONS RESEARCH & MANAGEMENT SCIENCE
Frederick S. Hillier, Series Editor, *Stanford University*

Saigal / *A MODERN APPROACH TO LINEAR PROGRAMMING*
Nagurney / *PROJECTED DYNAMICAL SYSTEMS & VARIATIONAL INEQUALITIES WITH APPLICATIONS*
Padberg & Rijai / *LOCATION, SCHEDULING, DESIGN AND INTEGER PROGRAMMING*
Vanderbei / *LINEAR PROGRAMMING*
Jaiswal / *MILITARY OPERATIONS RESEARCH*
Gal & Greenberg / *ADVANCES IN SENSITIVITY ANALYSIS & PARAMETRIC PROGRAMMING*
Prabhu / *FOUNDATIONS OF QUEUEING THEORY*
Fang, Rajasekera & Tsao / *ENTROPY OPTIMIZATION & MATHEMATICAL PROGRAMMING*
Yu / *OR IN THE AIRLINE INDUSTRY*
Ho & Tang / *PRODUCT VARIETY MANAGEMENT*
El-Taha & Stidham / *SAMPLE-PATH ANALYSIS OF QUEUEING SYSTEMS*
Miettinen / *NONLINEAR MULTIOBJECTIVE OPTIMIZATION*
Chao & Huntington / *DESIGNING COMPETITIVE ELECTRICITY MARKETS*
Weglarz / *PROJECT SCHEDULING: RECENT TRENDS & RESULTS*
Sanin & Polatoglu / *QUALITY, WARRANTY AND PREVENTIVE MAINTENANCE*
Tavares / *ADVANCES MODELS FOR PROJECT MANAGEMENT*
Tayur, Ganeshan & Magazine / *QUANTITATIVE MODELS FOR SUPPLY CHAIN MANAGEMENT*
Weyant, J. / *ENERGY AND ENVIRONMENTAL POLICY MODELING*
Shanthikumar, J.G. & Sumita, U. / *UNAPPLIED PROBABILITY AND STOCHASTIC PROCESSES*
Liu, B. & Esogbue, A.O. / *DECISION CRITERIA AND OPTIMAL INVENTORY PROCESSES*
Gal, T., Stewart, T.J., Hanne, T. / *MULTICRITERIA DECISION MAKING: Advances in MCDM Models, Algorithms, Theory, and Applications*
Fox, B.L. / *STRATEGIES FOR QUASI-MONTE CARLO*
Hall, R.W. / *HANDBOOK OF TRANSPORTATION SCIENCE*
Grassman, W.K. / *COMPUTATIONAL PROBABILITY*
Pomerol, J.-C. & Barba-Romero, S. / *MULTICRITERION DECISION IN MANAGEMENT*
Axsater, S. / *INVENTORY CONTROL*
Wolkowicz, H., Saigai, R., & Vandenberghe, L. / *HANDBOOK OF SEMI-DEFINITE PROGRAMMING: Theory, Algorithms, and Applications*
Hobbs, B.F. & Meier, P. / *ENERGY DECISIONS AND THE ENVIRONMENT: A Guide to the Use of Multicriteria Methods*
Dar-El, E. / *HUMAN LEARNING: From Learning Curves to Learning Organizations*
Armstrong, J.S. / *PRINCIPLES OF FORECASTING: A Handbook for Researchers and Practitioners*
Baisamo, S., Personé, V. & Onvural, R. / *ANALYSIS OF QUEUEING NETWORKS WITH BLOCKING*
Bouyssou, D. et al. / *EVALUATION AND DECISION MODELS: A Critical Perspective*
Hanne, T. / *INTELLIGENT STRATEGIES FOR META MULTIPLE CRITERIA DECISION MAKING*
Saaty, T. & Vargas, L. / *MODELS, METHODS, CONCEPTS and APPLICATIONS OF THE ANALYTIC HIERARCHY PROCESS*
Chatterjee, K. & Samuelson, W. / *GAME THEORY AND BUSINESS APPLICATIONS*
Hobbs, B. et al. / *THE NEXT GENERATION OF ELECTRIC POWER UNIT COMMITMENT MODELS*
Vanderbei, R.J. / *LINEAR PROGRAMMING: Foundations and Extensions, 2nd Ed.*
Kimms, A. / *MATHEMATICAL PROGRAMMING AND FINANCIAL OBJECTIVES FOR SCHEDULING PROJECTS*
Baptiste, P., Le Pape, C. & Nuijten, W. / *CONSTRAINT-BASED SCHEDULING*
Feinberg, E. & Shwartz, A. / *HANDBOOK OF MARKOV DECISION PROCESSES: Methods and Applications*

**** A list of the more recent publications in the series is at the front of the book ****